T0257250

Mechanical Properties and Performance of Engineering Ceramics and Composites VIII

Mechanical Properties and Performance of Engineering Ceramics and Composites VIII

A Collection of Papers Presented at the 37th International Conference on Advanced Ceramics and Composites January 27–February 1, 2013 Daytona Beach, Florida

Edited by
Dileep Singh
Jonathan Salem

Volume Editors
Soshu Kirihara
Sujanto Widjaja

The American Ceramic Society

WILEY

Published by John Wiley & Sons, Inc., Hoboken, New Jersey.
Published simultaneously in Canada.

For general information on our other products and services or for technical support, please contact our
Customer Care Department within the United States at (800) 762-2974, outside the United States at
(317) 572-3993 or fax (317) 572-4002.

Wiley also publishes its books in a variety of electronic formats. Some content that appears in print may
not be available in electronic formats. For more information about Wiley products, visit our web site at
www.wiley.com.

Library of Congress Cataloging-in-Publication Data is available.

ISBN: 978-1-118-80747-7
ISSN: 0196-6219

Printed in the United States of America.

10 9 8 7 6 5 4 3 2 1

Contents

PROCESSING AND PROPERTIES OF FIBERS AND CERAMICS

Preface

This volume is a compilation of papers presented in the Mechanical Behavior and Performance of Ceramics & Composites symposium during the 37th International Conference & Exposition on Advanced Ceramics and Composites (ICACC) held January 27 to February 1, 2013, in Daytona Beach, Florida.

This long-standing symposium received presentations on a wide variety of topics thus providing the opportunity for researchers in different areas of related fields to interact. This volume emphasizes some practical aspects of real-world engineering applications of materials such as oxidation, fatigue, wear, nondestructive evaluation, and mechanical behavior as associated with systems ranging from rare earth aluminates to multilayer composite armor. Symposium topics included:

- Composites: Fibers, Interfaces Modeling and Applications
- Fracture Mechanics, Modeling, and Mechanical Testing
- Nondestructive Evaluation
- Processing-Microstructure-Properties Correlations
- Tribological Properties of Ceramics and Composites

Significant time and effort is required to organize a symposium and publish a proceeding volume. We would like to extend our sincere thanks and appreciation to the symposium organizers, invited speakers, session chairs, presenters, manuscript reviewers, and conference attendees for their enthusiastic participation and contributions. Finally, credit also goes to the dedicated, tireless and courteous staff at The American Ceramic Society for making this symposium a huge success.

DILEEP SINGH
Argonne National Laboratory

JONATHAN SALEM
NASA Glenn Research Center

Introduction

This issue of the Ceramic Engineering and Science Proceedings (CESP) is one of nine issues that has been published based on manuscripts submitted and approved for the proceedings of the 37th International Conference on Advanced Ceramics and Composites (ICACC), held January 27–February 1, 2013 in Daytona Beach, Florida. ICACC is the most prominent international meeting in the area of advanced structural, functional, and nanoscopic ceramics, composites, and other emerging ceramic materials and technologies. This prestigious conference has been organized by The American Ceramic Society's (ACerS) Engineering Ceramics Division (ECD) since 1977.

The 37th ICACC hosted more than 1,000 attendees from 40 countries and approximately 800 presentations. The topics ranged from ceramic nanomaterials to structural reliability of ceramic components which demonstrated the linkage between materials science developments at the atomic level and macro level structural applications. Papers addressed material, model, and component development and investigated the interrelations between the processing, properties, and microstructure of ceramic materials.

The conference was organized into the following 19 symposia and sessions:

Symposium 1	Mechanical Behavior and Performance of Ceramics and Composites
Symposium 2	Advanced Ceramic Coatings for Structural, Environmental, and Functional Applications
Symposium 3	10th International Symposium on Solid Oxide Fuel Cells (SOFC): Materials, Science, and Technology
Symposium 4	Armor Ceramics
Symposium 5	Next Generation Bioceramics
Symposium 6	International Symposium on Ceramics for Electric Energy Generation, Storage, and Distribution
Symposium 7	7th International Symposium on Nanostructured Materials and Nanocomposites: Development and Applications

Symposium 8	7th International Symposium on Advanced Processing & Manufacturing Technologies for Structural & Multifunctional Materials and Systems (APMT)
Symposium 9	Porous Ceramics: Novel Developments and Applications
Symposium 10	Virtual Materials (Computational) Design and Ceramic Genome
Symposium 11	Next Generation Technologies for Innovative Surface Coatings
Symposium 12	Materials for Extreme Environments: Ultrahigh Temperature Ceramics (UHTCs) and Nanolaminated Ternary Carbides and Nitrides (MAX Phases)
Symposium 13	Advanced Ceramics and Composites for Sustainable Nuclear Energy and Fusion Energy
Focused Session 1	Geopolymers and Chemically Bonded Ceramics
Focused Session 2	Thermal Management Materials and Technologies
Focused Session 3	Nanomaterials for Sensing Applications
Focused Session 4	Advanced Ceramic Materials and Processing for Photonics and Energy
Special Session	Engineering Ceramics Summit of the Americas
Special Session	2nd Global Young Investigators Forum

The proceedings papers from this conference are published in the below nine issues of the 2013 CESP; Volume 34, Issues 2–10:

- Mechanical Properties and Performance of Engineering Ceramics and Composites VIII, CESP Volume 34, Issue 2 (includes papers from Symposium 1)
- Advanced Ceramic Coatings and Materials for Extreme Environments III, Volume 34, Issue 3 (includes papers from Symposia 2 and 11)
- Advances in Solid Oxide Fuel Cells IX, CESP Volume 34, Issue 4 (includes papers from Symposium 3)
- Advances in Ceramic Armor IX, CESP Volume 34, Issue 5 (includes papers from Symposium 4)
- Advances in Bioceramics and Porous Ceramics VI, CESP Volume 34, Issue 6 (includes papers from Symposia 5 and 9)
- Nanostructured Materials and Nanotechnology VII, CESP Volume 34, Issue 7 (includes papers from Symposium 7 and FS3)
- Advanced Processing and Manufacturing Technologies for Structural and Multi functional Materials VII, CESP Volume 34, Issue 8 (includes papers from Symposium 8)
- Ceramic Materials for Energy Applications III, CESP Volume 34, Issue 9 (includes papers from Symposia 6, 13, and FS4)
- Developments in Strategic Materials and Computational Design IV, CESP Volume 34, Issue 10 (includes papers from Symposium 10 and 12 and from Focused Sessions 1 and 2)

The organization of the Daytona Beach meeting and the publication of these proceedings were possible thanks to the professional staff of ACerS and the tireless dedication of many ECD members. We would especially like to express our sincere thanks to the symposia organizers, session chairs, presenters and conference attendees, for their efforts and enthusiastic participation in the vibrant and cutting-edge conference.

ACerS and the ECD invite you to attend the 38th International Conference on Advanced Ceramics and Composites (http://www.ceramics.org/daytona2014) January 26-31, 2014 in Daytona Beach, Florida.

To purchase additional CESP issues as well as other ceramic publications, visit the ACerS-Wiley Publications home page at www.wiley.com/go/ceramics.

SOSHU KIRIHARA, *Osaka University, Japan*
SUJANTO WIDJAJA, *Corning Incorporated, USA*

Volume Editors
August 2013

Characterization and Modeling of Ceramic Matrix Composites

ACOUSTIC EMISSION AND ELECTRICAL RESISTIVITY MONITORING OF SIC/SIC COMPOSITE CYCLIC BEHAVIOR

Christopher R. Baker and Gregory N. Morscher

Mechanical Engineering Department, The University of Akron, Akron, OH

1. Abstract

Successful implementation of SiC/SiC composites in rotating components demands an excellent understanding of their cyclic stress and fatigue behavior. Two fiber-reinforced melt-infiltrated composites were tested under several cyclic stress conditions. In situ acoustic emission and electrical resistance was used to monitor damage accumulation and understand time and stress dependant degradation of mechanical properties. Post test microscopy and fractography was then used to observe microstructural anomalies which suggest possible damage progression mechanisms. Electrical resistance measurements were in good agreement with decreasing interfacial shear stress and show potential as a monitoring technique for fatigue related damage.

2. Introduction

Ceramic matrix composites (CMCs) are being investigated for potential use in the hot section of aerospace and power generating turbines. Their exceptional thermal stability at elevated temperatures (~1200°C) makes them uniquely able to allow current hot-section temperatures to be raised. However, should CMCs ever be able to be implemented in rotating components, an excellent understanding of fatigue performance and damage accumulation is necessary. This is made more difficult in the presence of damage (e.g. matrix cracking, fiber pull-out, etc.). The mechanics of loading, unloading and reloading after damage have initiated have been examined in the literature[1,2]. Similarly, cyclic loading effects on the interphase have been investigated and are an important part in the mechanical behavior of a damaged CMC[3].

Non-destructive techniques have been investigated to better understand the damage initiation, progression and intensity in many CMC systems[3-5]. These include acoustic emission (AE), electrical resistance (ER), flash thermography and acousto-ultrasonics. ER techniques are particularly novel and a cohesive mechanical-electrical model has not yet been developed, although a few have been suggested. The models suggested typically rely on factors such as pristine material resistivity, number of cracks per unit distance, stress, etc.

In this work, two different materials were loaded cyclically in different loading parameters. With the aim of understanding the physical nature of the damage accumulation through non-destructive health-monitoring techniques, each test was monitored using Modal Acoustic Emission and ER. Comparison of the pristine mechanical behavior and cyclically loaded material through these measurements, one may be better able to understand the true nature of the damage accumulation in CMCs.

3

3. Experimental

Two balanced 0°/90° 5-harness-satin liquid silicon infiltrated composites were tested in varying cycling conditions. All composites are slurry cast SiC/SiC composites each with a different fiber type: Sylramic-iBN[*] and Tyranno ZMI[**]. Fiber volume fractions were determined by dividing the cross-sectional area of the composite by the fiber area (determined by multiplying fiber area, number of fibers per tow, number of plies and tows per ply.) All composites were manufactured by Goodrich[***] and machined into dog-bone coupons. Mechanical, physical and geometric properties are listed in Table I. Coupons were loaded/unloaded at a rate of 6kN/min. The ZMI reinforced composite underwent 110 total cycles and the Sylramic-iBN composite was loaded/unloaded 250 cycles.

Strain was measured using an 1" (25.4mm) extensometer with ±2% travel on an MTS self-aligning system with hydraulic pressure grips. ER was measured using an Agilent[****] 34420A digital multimeter over the gage length of 25.4 mm. AE signals were acquired using a Digital Wave[*****] Fracture Wave Detector. Wide-band sensors (50kHz-2MHz) were used and a threshold crossing technique was used to determine lengthwise location of events, similar to a technique used by Morscher[6]. Only events occurring in the gage section (middle 25.4 mm) were used in the analysis. Figure 1 shows the typical layout for a coupon prior to testing.

Table I. List of physical and mechanical properties of each composite

Specimen	E (GPa)	AE Onset (MPa)*	Gage Width (mm)	Thickness (mm)	Pristine Resistivity (Ω-mm)	Residual Stress (MPa)	Fiber Volume Fraction
ZMI	171	68	10.2	4.04	0.498	12	0.26
Sylramic-iBN	201	140	10.27	2.35	0.648	-60	0.36

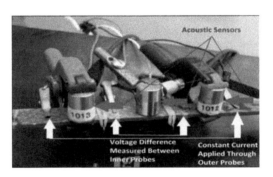

Figure 1. Experimental layout of a CMC prior to testing

4. Results and Discussion

Stress-strain behavior is shown in Figure 2 for each of the composites. The composites were loaded for several cycles under a particular stress, then loaded to a higher stress for several more cycles. This pattern continued, as shown in Figure 2. Note that none of the composites were loaded to failure. The higher strains accumulated in the ZMI composite are due to the lower fiber modulus compared to the Sylramic-iBN. Interestingly, strain seems to accumulate both at the zero stress condition and the fully loaded condition more with each cycle. This is consistent with both samples, and the interphase degradation is likely the reason for such behavior.

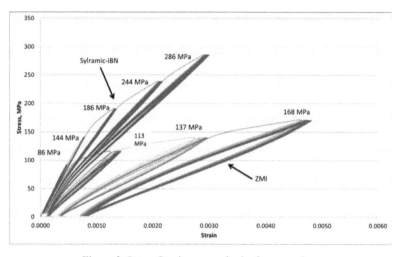

Figure 2. Stress Strain curves for both composites.

4.1. Acoustic Emission and Damage Accumulation

Acoustic emission has been shown to be effective at quantifying matrix cracking in various CMCs[7]. In particular, the cumulative AE energy (summation of the energies of each successive event) has been shown to have a direct correlation with stress-dependent crack density because high energy events, which dominate the cumulative energy curve, correspond to large matrix cracks. The acoustic and electrical behavior for each composite is shown in Figure 3. Figure 3a and 3b show that the most of the acoustic energy, i.e., matrix crack formation, occurs during the first cycle loading at each stress level. For the ZMI composites, high energy events occurred on the second and third cycles at the lowest stress condition (Figure 3). However, after the composite reached the crack saturation stress (~ 140 MPa), no further significant high energy events were detected. For the Sylramic-iBN composite (Figure 3b), high energy events did not occur after the first stress cycle.

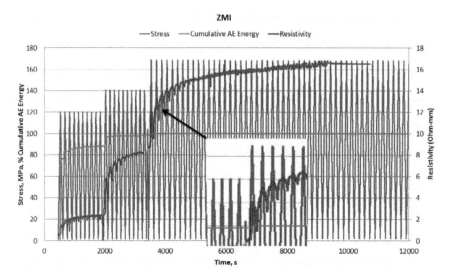

Figure 3a. ZMI composite AE and ER behavior

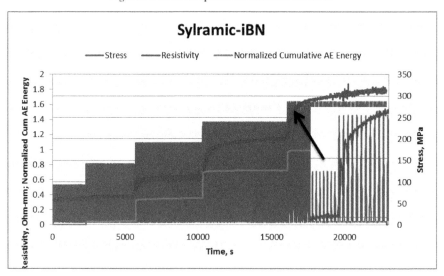

Figure 3b. Sylramic-iBN composite AE and ER behavior

4.2. Interfacial Shear Stress Degradation

Rouby and Reynaud[3] proposed an exponential decay of the interfacial shear stress, τ, as a function of cycles. This is bounded by a lower limiting value, τ_∞. It takes the following form:

$$\tau = \tau_0 N^{-t} \quad (N < N_\infty)$$

$$\tau = \tau_\infty \quad (N > N_\infty) \tag{1}$$

Where, for each composite system, τ_0, t and τ_∞ are unique parameters. In this work, τ was calculated for each cycle by combining equations (3) and (7) in Curtin[2] and is given by:

$$\tau = \frac{\alpha^2 r \sigma_t^2 \rho_c}{2\left(\epsilon - \frac{\sigma}{E_c}\right)E_f} \tag{2}$$

Where:

$$\alpha = \frac{(1-f)E_m}{fE_c} \tag{3}$$

and r is the fiber radius, σ_t is the total stress (applied and residual), ρ_c is the crack density. Fiber volume fraction used in the equation is only the fiber volume fraction in the loading direction (i.e. half the total fiber volume fraction.) Crack density was found using the technique of Morscher[7], wherein a fraction of total accumulated AE energy is normalized by total cumulative AE energy, then multiplied by the saturation crack density, giving a stress-dependent crack density. Saturation crack densities were found from post-test microscopy of similar materials. Matrix modulus was calculated using a rule of mixtures approach, that is:

$$E_m = \frac{E_c - fE_f}{1-f} \tag{4}$$

The values of τ are shown in Figure 4. Curve fitting parameters adopted from the model in (1) are also shown. Note that τ values for the first 50 cycles of the Sylramic-iBN coupon resulted in no cracking, so the chart considers the first cycle at 141 MPa to be the first cycle. the power-law fits the data reasonably well.

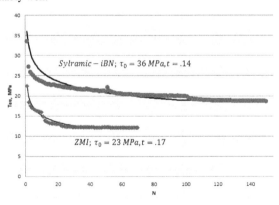

Figure 4. ISS cyclic degradation

4.3. Electrical Resistance

After the first cycle at a particular stress, the acoustic activity is negligible, although the ER continues to increase. Looking again at Figure 2, we can see that strain continues to accumulate. This suggests that what controls increasing ER is what causes increasing strain accumulation, which is most likely the continued wear of the interphase, responsible for the decrease in ISS shown in Figure 3. Also, cycling in the elastic region of the Sylramic-iBN composite yielded no significant change in resistivity, indicating that ER is more sensitive to damage than due to a stress response of the material.

Figure 5 shows conductivity (the reciprocal of the resistivity) and the decrease in ISS due to cyclic loading for the ZMI and Sylramic-iBN composites under the highest stress condition (ZMI) and intermediate stress condition (Sylramic-iBN). The two properties decrease very similarly with increasing number of cycles. Also, in the initial cycles of each composite, there is a recovery, to some degree, of the electrical resistance to a lower value upon unloading (see inset in Figure 3a). As cycling continues, this ER recovery becomes smaller and eventually becomes completely negligible. Since the matrix is generally much more conductive than the fiber, interphase or CVI SiC (due to the presence of continuous Si), the resistivity of the composite largely relies on the ability of the matrix to conduct and transfer electric current. As cycling progresses, conductivity of the composite decreases, and is likely associated with the increasing difficulty of current transfer to the matrix in a matrix segment between matrix cracks. A possible mechanism is the cracks not fully closing on unloading, but this needs further investigation.

Observing the inlay of Figure 3a, the ER consistently recovers some conductivity upon unloading, and increases to new maximum value each cycle on loading. Because there is very little new acoustic activity after the first cycle, we can deduce that the increase in resistivity isn't due to new matrix crack formation, but rather an increase in fiber-matrix contact resistance, which is, in turn, associated with reduced ISS, shown in Figure 5. Presumably, interfacial wear reduces fiber-to-matrix contact and less amount of current carried by the matrix.

After the composite was fully cycled, monotonic tensile loading to failure was performed on the Sylramic-iBN to compare fracture surfaces to a non-cycled monotonic test of the same material. Post-test micrographs are shown in Figure 6. It can be easily seen that the degree of fiber pull-out is more significant in the post-cycled composite than the monotonic to-failure composite. The larger scale if fiber pull-out indicates that there is, in fact, a degradation of the ISS due to cyclic loading.

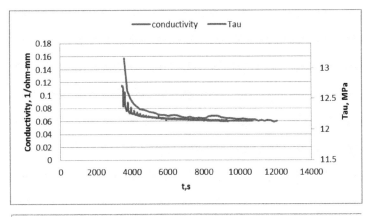

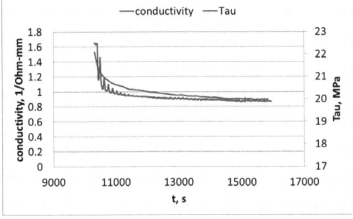

Figure 5. Conductivity-ISS relationship; ZMI (top) and Sylramic-iBN (bottom)

5. Conclusion

It has been shown that there is a reduction of the interfacial shear stress due to cyclic loading, indicated by accumulated strain in the composite, fracture surface comparison and increased electrical resistivity. A power-law model of the decreasing ISS is appropriate for these composites, and may be appropriate for the ER accumulation due to cycles. Furthermore, there appears to be a direct correlation between τ and the electrical conductivity of the composite. Future modeling of damage in composites should contain changing interphase condition as a function of loading cycles, whether hysteretic or constant load.

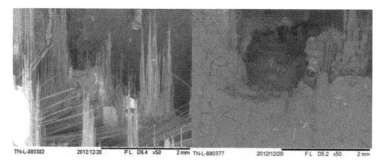

Figure 6. Figure showing fracture surface of cycled Sylramic-iBN composite (left) and non-cycled Sylramic-iBN Composite (right)

6. References

1. Pryce, a. W., & Smith, P. a. (1993). Matrix cracking in unidirectional ceramic matrix composites under quasi-static and cyclic loading. *Acta Metallurgica et Materialia*, *41*(4), 1269–1281. doi:10.1016/0956-7151(93)90178-U

2. Curtin, W. A., Ahn, B. K., & Takeda, N. (1998). Modeling Brittle and Tough Stress-Strain Behavior in Unidirectional Ceramic Matrix Composites, *46*(10), 3409–3420.

3. Rouby, D., & Reynaud, P. (1993). Modification During Load Cycling in Ceramic-Matrix Fiber Composites, *48*, 109–118.

4. Smith, C., Morscher, G., & Xia, Z. (2008). Monitoring damage accumulation in ceramic matrix composites using electrical resistivity. *Scripta Materialia*, *59*(4), 463–466. doi:10.1016/j.scriptamat.2008.04.033

5. Morscher, G. N., & Baker, C. (2012). GT2012-69167 Use of Electrical Resistivity and Acoustic Emission to Assess Impact Damage States in Two SiC-Based CMCs. *ASME Turbo Expo 2012*, 1–6.

6. Morscher, G. N. (1999). Modal acoustic emission of damage accumulation in a woven SiC / SiC composite. *Composites Science and Technology*, *59*, 687–697.

7. Morscher, G. N. (2004). Stress-dependent matrix cracking in 2D woven SiC-fiber reinforced melt-infiltrated SiC matrix composites. *Composites Science and Technology*, *64*(9), 1311–1319. doi:10.1016/j.compscitech.2003.10.022

* Dow Corning, Inc., Midland MI
** Ube Industries, Japan
*** Brecksville, OH
**** Santa Clara, CA
***** Centennial, CO

CHARACTERIZATION OF SIC/SICN CERAMIC MATRIX COMPOSITES WITH MONAZITE FIBER COATING

Enrico Klatt[1], Klemens Kelm[2], Martin Frieß[1], Dietmar Koch[1], and Heinz Voggenreiter[1]

[1]Institute of Structure and Design, German Aerospace Center (DLR), Stuttgart, Germany
[2]Materials Research Institute, German Aerospace Center (DLR), Cologne, Germany

ABSTRACT

Non-oxide ceramic matrix composites (CMCs) based on silicon carbide fibers (Tyranno SA3; UBE Industries Ltd., Japan) were manufactured via PIP-process using a polysilazane precursor. Prior to PIP-process, SiC-fabrics (two fabric types: PSA-S17I16PX and PSA-S17F08PX) were coated with oxidation resistant monazite (LaPO$_4$) which was derived from rhabdophane solution. Two different monazite coating processes were used: Foulard coating (FC) and dip coating (DC). Several iterations have been performed to reach the final coating thickness of ~100 nm and ~300 nm for FC and DC, respectively. Evolution of monazite coating for FC/DC was determined by calculating theoretical coating thickness and scanning electron microscopy (SEM) and differences will be discussed. After PIP-process, CMC materials were mechanically (3pt.-bending test) and microstructurally (SEM, TEM) characterized before and after exposure to air (T=1100°C, 20 hr). Before exposure, CMC materials with homogeneous fiber coating exhibit moderate strength (~150–170 MPa) and damage tolerant behavior with significant fiber-pullout. After exposure, strength degradation of only 16%, less damage tolerant behavior and less fiber-pullout could be observed. Reasons for degradation might be silica formation which was observed on fibers and matrix as well as aluminum (incorporated in Tyranno SA3-fiber as a sintering aid) which diffused into fiber coating. Possible oxidation mechanisms will be presented and discussed.

INTRODUCTION

Non-oxide CMCs like SiC/SiC materials exhibit very high specific strength and excellent high temperature resistance. Nevertheless, one of the outstanding disadvantages is the intrinsic lack of oxidation resistance. The application of oxidation resistant fiber coatings is of interest to improve oxidation resistance of non-oxide CMCs. Monazite (LaPO$_4$) is an attractive candidate for fiber coating due to its oxidation resistance, high melting point (2072°C),[1] low hardness (Mohs hardness 5.5),[2] and high thermodynamic stability towards most of common ceramics. Morgan et al. demonstrated for the first time the functionality of monazite coating (weak fiber/matrix bonding which enables crack deflection and fiber-pullout) on sapphire fibers embedded in a ceramic matrix.[2] Further investigations showed planar deformation of monazite between different gliding planes which panders to fiber-pullout.[3,4] Improved damage tolerance as well as significant fiber-pullout could be demonstrated in oxide CMCs before and after exposure to air.[5-7] Monazite as fiber coating in silicon carbide fiber reinforced CMCs has not been published to our knowledge.

There are different methods to deposit monazite on fibers: Next to gas phase deposition (PVD or CVD),[8] sol-gel-processes,[9,10] or electrophoretic deposition (EPD)[11] in-situ deposition based on wet chemical dip processes represents a promising method to coat fibers with monazite. Air Force Research Laboratory (AFRL, Ohio, USA) developed a dip coating process to deposit monazite via heterogeneous nucleation which is described in detail elsewhere.[10,12-17] This coating process was initially applied to Nextel™-fibers (N610/720, 3M, St. Paul, USA)[12,13,18,19] and later transferred to SiC-fibers.[16,20] Boakye et al. found that monazite in contact with SiC or SiC+SiO$_2$ is not stable in reducing atmospheres and forms lanthanum silicate:[16]

$$2LaPO_{4(s)} + 2SiC_{(s)} \rightarrow La_2Si_2O_{7(s)} + CO_{(g)} + C_{(f)} + P_{2(g)} \tag{1}$$
$$6LaPO_{4(s)} + 2SiC_{(s)} + SiO_{2(s)} \rightarrow 3La_2Si_2O_{7(s)} + 5CO_{(g)} + 3P_{2(g)} \tag{2}$$

Cinibulk et al. found a correlation between oxygen partial pressure and reduction of monazite by SiC for temperatures of 1200–1400°C.[21] For oxygen partial pressures <0.5 Pa, monazite reduces at 1200°C according to Eq. 1. Mawdsley et al. observed that monazite in contact with carbon reacts to lanthanum-oxyphosphate in oxidizing atmospheres at temperatures >750°C:[22]

$$3LaPO_{4(s)} + 5C_{(s)} \rightarrow La_3PO_{7(s)} + 5CO_{(g)} + P_{2(g)} \tag{3}$$

These reactions have to be considered for the choice of fiber/matrix configuration, CMC manufacturing process as well as process conditions for CMC development. The objective of this study is to apply monazite as a fiber coating for SiC-fibers and demonstrate monazite as a functional and effective interface in SiC/SiCN materials manufactured via polymer infiltration and pyrolysis process for the first time.

EXPERIMENTAL

As fiber reinforcement, two different types of plain weave fabrics (PSA-S17I16PX and PSA-S17F08PX) based on commercially available SiC-fiber type Tyranno SA3™ (UBE Industries Ltd., Yamaguchi, Japan) were used (Table 1).

For monazite coating, rhabdophane solution has to be prepared: The solution is based on a mixture of lanthanum(III)-nitrate-hexahydrate (La(NO$_3$)$_3$ x 6H$_2$O, 99.9%, Alfa Aesar GmbH &Co.KG, Karlsruhe, Germany), orthophosphoric acid (H$_3$PO$_4$, 85%, Alfa Aesar), citric acid monohydrate (C$_6$H$_8$O$_7$ x H$_2$O, Carl Roth GmbH&Co.KG, Karlsruhe, Germany) and deionized water. The molar ratio of La:P was adjusted to 1:1 (stoichiometric) and of La:Cit to 1:2.5. The concentration of the solution was about 36 g LaPO$_4$/liter. To prevent precipitation of rhabdo-phane before coating process, solution has to be cooled to ~5°C. Prior to monazite coating process, fabrics were cut (300x90 mm) and desized (600°C, 10 min, air). After desizing, fabrics were coated with monazite by using two different coating processes: Foulard coating (FC, Figure 1 a) and dip coating (DC, Figure 1 b). A foulard is a machine, where two counter-rotating (speed: ~3 m/min) rubber rolls were pressed together (~1 bar) and fabrics will be fed through them.

Due to the fact, that theoretical monazite coating thickness (calculated by mass gain) via FC is less than via DC, two coating and drying cycles have been performed for FC before intercoat firing (ICF). Several coating cycles (coating+drying+ICF) can be performed to increase coating thickness. FC is much more time consuming compared to DC (about factor 3). There-fore, the target of coating thickness for FC and DC was 100 nm and 300 nm, respectively.

Table 1. Properties of Tyranno SA3™ fabrics (UBE Industries Ltd., Japan).[23,24]

Fiber		Tyranno SA3™	
Weave specification	-	PSA-S17I16PX	PSA-S17F08PX
Weave specification (abbrev.)	-	SA3-I16	SA3-F08
Weave type	-	Plain weave 1/1	Plain weave 1/1
Weight of area	g/m²	260	240
Filament diameter	μm	7.5	10.0
Filaments/roving	-	1600	800
Roughness	nm	~200	~200

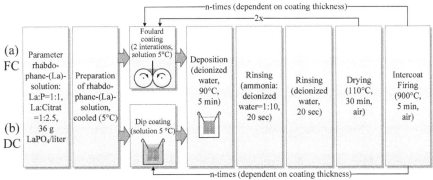

Figure 1.   Schematic monazite fiber-coating processes: (a) Foulard coating (FC), (b) Dip coating (DC).

For CMC manufacturing, polymer infiltration and pyrolysis process (PIP) will be used as described elsewhere.[25] A PIP-cycle consists of polymer infiltration (polysilazane PSZ20, Clariant Advanced Materials GmbH, Sulzbach, Germany) and drying using resin transfer molding process (RTM) as well as subsequent pyrolysis (T_{max}=1000°C, nitrogen). The fiber volume content was adjusted to ~45 vol.%. To decrease porosity (<10 vol.%), in total six PIP-cycles have been performed. Finally, CMC-materials were mechanically (3pt.-bending test according to DIN EN 658-3; universal testing machine, Zwick/Roell, Ulm, Germany) as well as micro-structurally (SEM: Ultra 55 Plus, Carl Zeiss NTS GmbH, Oberkochen, Germany; TEM: Tecnai F30, Philips, Eindhoven, the Netherlands) characterized before and after exposure to air (1100°C, 20 hr).

RESULTS AND DISCUSSION

Monazite coating processes: Foulard coating (FC) and dip coating (DC)

EDS analysis (by SEM) of monazite (dehydrated rhabdophane (900°C, 5 min, air)) which was deposited by heterogeneous nucleation from rhabdophane solution shows a good agreement with theoretical composition of $LaPO_4$ (Figure 2 a). Nevertheless, monazite is slightly P-rich which is consistent to literature given elsewhere.[10] Carbon with a concentration of ~5 at.% could be observed, which might be caused by pyrolysis of residual citric acid (from rhabdophane solution) due to insufficient rinsing. XRD pattern of monazite correlates very well with peaks (from literature) for crystalline $LaPO_4$ (Figure 2 b). Nevertheless, the distinctive background might be caused by small amount of amorphous phases.

To compare the evolution of monazite coating thickness of FC and DC, theoretical mo-nazite coating thickness was calculated from mass gain after each coating cycle (Figure 3 a, b). Evolution of coating thickness is similar for FC and DC (~10–15 nm for each coating cycle), even though one FC cycle consists of two cycles of saturation and drying before ICF compared to one cycle for DC (ref. Figure 1). In total 5–7 and 14–15 coating cycles (dependent on fiber diameter) are required for FC (target: 100 nm) and DC (target: 300 nm), respectively. The bigger the fiber diameter (SA3-F08: 10.0 μm; SA3-I16: 7.5 μm; ref. Table 1), the thicker the deposited monazite coating (for FC and DC) which is mainly caused by decreasing specific surface with increasing fiber diameter.

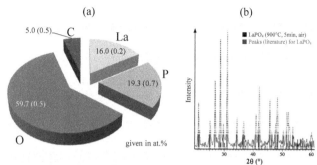

Figure 2. Analysis of monazite deposited via heterogeneous nucleation and dehydrated at 900°C (5 min, air): (a) EDS analysis by SEM, (b) XRD powder diffractogram.

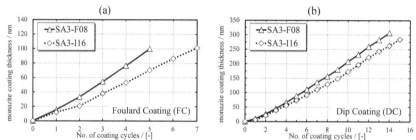

Figure 3. Development of theoretical monazite coating thickness: (a) Foulard coating (target: ~100 nm), (b) Dip coating (target: ~300 nm).

Figure 4 shows microstructure of monazite coated (via FC and DC) SA3-I16 fabrics after one coating cycle. Monazite coating is porous for both, FC and DC. For FC, the saturation quality is much better than for DC leading to better covering of fibers, especially in inner area (inner bundle) of the fabrics. Good saturation quality for FC may be due to counter-rotating rubber rolls (pressed together) enabling better immersion of rhabdophane solution than DC.

This leads to better homogeneity (with respect to coating thickness and covering ratio) of monazite coating for FC compared to DC (Figure 5). For FC, fibers in outer ("loose" packing) as well as inner area ("tight" packing) of fabric layers are well covered with monazite with a homogeneous coating thickness. No bridging of monazite coating could be observed. Nevertheless, fibers in regions with "loose" packing exhibit thicker monazite coating compared

Figure 4. Comparison of monazite coating for FC and DC after one coating cycle for Tyranno SA3-I16.

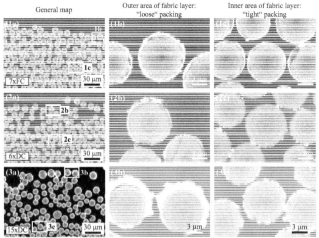

Figure 5. Comparison of formation of monazite coating for (1 a-c) FC and (2, 3 a-c) DC in outer and inner area of fabric layer (fabric type: SA3-I16).

to regions with "tight" packing (approximately factor 2). Compared to FC, DC shows inhomogeneous fiber coating where filaments are not fully covered with monazite (especially in regions with "tight" packing). Furthermore, with increasing numbers of coating cycles, the difference in monazite coating thickness between "loose" and "tight" packing increases. This leads to local bridging of monazite coating in regions with "loose" packing which limits immersion of rhabdophane in subsequent coating cycles and promotes increasing gradient in coating thickness.

Mechanical characterization of SiC_{LaPO4}/SiCN composites

After six PIP-cycles, SiC_{LaPO4}/SiCN materials exhibit low open porosity (~6.3 vol.%) and a density of ~2.5 g/cm^3 (measured by Archimedes method). Typical stress-strain-curves from 3pt.-bending test before and after exposure to air (1100°C, 20 hr) for different fabric types (SA3-I16, SA3-F08) and monazite coating processes (FC, DC) are given in Figure 6.

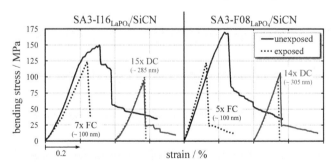

Figure 6. Typical stress-strain-curves (from 3pt.-bending test) for SiC_{LaPO4}/SiCN before and after exposure to air (1100°C, 20 hr) with different fabric types (SA3-I16 and SA3-F08) and monazite fiber coating deposited by different methods (FC and DC).

Before exposure to air, SiC_{LaPO4}/SiCN (FC) is characterized by bending strengths of ~150–170 MPa, distinct non-linear behavior and no abrupt failure after reaching bending strength which indicates damage tolerant behavior. Compared to that, SiC_{LaPO4}/SiCN (DC) exhibits lower bending strength (~90–100 MPa), less non-linear behavior and abrupt failure after reaching bending strength (brittle material behavior), even there is a much thicker theoretical monazite coating (300 nm compared to 100 nm for FC). Differences in fracture behavior and bending strength are mainly caused by differences in fiber/matrix-bonding due to differences in homogeneity in monazite fiber coating (ref. Figure 5) and fiber strength degradation during coating process, which will be discussed later in this study. After exposure to air, all materials show embrittlement (less non-linear behavior, mainly abrupt failure after reaching bending strength). Bending strength of SiC_{LaPO4}/SiCN (FC) drops to ~125–130 MPa whereas for SiC_{LaPO4}/SiCN (DC) no significant changes can be observed. Strength loss of only ~16% for SiC_{LaPO4}/SiCN (FC) is very low compared to similar CMC materials with pyrolytic carbon (pyc) interface, where the strength loss was ~70% for same exposure conditions.[25] We attribute this to the oxidation resistant monazite coating compared to pyc with low oxidation resistance. Statistics of mechanical characterization of SiC_{LaPO4}/SiCN before and after exposure to air are summarized in Figure 7.

Before exposure to air, the bending strength level of SiC_{LaPO4}/SiCN is much lower (approximately 50%) compared to SiC_{pyc}/SiCN manufactured via same PIP-process, where bending strength is ~300 MPa.[25] This might be caused by fiber strength degradation during monazite coating process. Hence, tensile tests have been performed with uncoated and monazite coated fiber tows (extracted from fabrics in warp direction). Independent of coating process (FC/DC), tow breaking load decreases with increasing number of coating cycles (equivalent to number of ICF) (Figure 8 a). The decrease is nearly linear as a function of coating cycles (Figure 8 b). After 15 coating cycles, fiber strength is reduced by ~50%. The reason for fiber strength degradation might be silica formation during ICF (especially due to formation of water vapor during dehydration of rhabdophane which accelerates silica formation[26-28]) and increasing coating thickness. Interestingly, fiber tow breaking load after 1st ICF (monazite coating) is higher than after 1st drying (rhabdophane coating). Rhabdophane has a lower strength level than SA3-fibers and is stronger bonded to SA3-fibers compared to porous monazite. Therefore, micro cracks initiated in rhabdophane interface at low stress levels can easily propagate into SA3-fibers leading to lower tow breaking loads. Summarizing, fiber strength degradation during monazite coating process is responsible for lower strength levels of SiC_{LaPO4}/SiCN compared to SiC_{pyc}/SiCN.

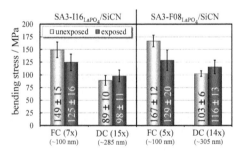

Figure 7.   Comparison of 3pt.-bending-strength (mean value and standard deviation=error bars) for SiC_{LaPO4}/SiCN (SA3-I16 and SA3-F08) with monazite fiber coating (FC and DC) before and after exposure to air (1100°C, 20 hr).

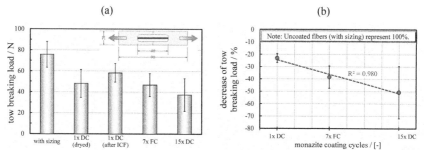

Figure 8.   (a) Tow breaking load for uncoated and monazite coated tows extracted from SA3-I16; (b) decrease of tow breaking load as a function of number of coating cycles.

Microstructural characterization of SiC$_{LaPO4}$/SiCN composites

Mechanical properties as well as fracture behavior of CMC materials are closely linked to microstructure. Fracture surfaces of SiC$_{LaPO4}$/SiCN (FC/DC) before and after exposure to air strongly correlate with mechanical behavior: Significant fiber-pullout (fiber-pullout lengths up to 1 mm) for SiC$_{LaPO4}$/SiCN (FC) before exposure to air (Figure 9 a) causes higher strength level, distinctive non-linear behavior and damage tolerance. In contrast to that, relatively smooth fracture surfaces with no significant fiber-pullout (fiber pullout lengths <20 μm) like for SiC$_{LaPO4}$/SiCN (DC) before exposure (Figure 9 c) or SiC$_{LaPO4}$/SiCN (FC/DC) after exposure (Figure 9 b, d) are strongly linked to low strength levels and an abrupt failure (brittle material behavior). The decrease of fiber-pullout for SiC$_{LaPO4}$/SiCN (FC) after exposure to air is caused by microstructural changes in monazite coating as well as SiC-fiber strength degradation due to silica formation.

Before exposure to air, monazite provokes a weak fiber/matrix bonding and enables fiber-pullout (fiber/matrix debonding in monazite like pullout-schlieren can be locally observed) (Figure 10 a). After exposure to air, no pullout-schlieren could be observed at all (Figure 10 b) which might be caused by changes in the interface. This will be discussed later in this study.

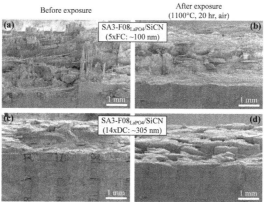

Figure 9.   SEM-images of fracture surfaces (3pt.-bending) of SA3-F08$_{LaPO4}$/SiCN with monazite fiber coating (5xFC: a, b; 14xDC: c, d) before and after exposure to air.

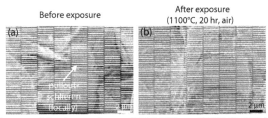

Figure 10. SEM-images of fracture surfaces of SA3-F08$_{LaPO4}$/SiCN (5xFC) before (a) and after (b) exposure to air.

Before exposure to air, fiber-pullout could be mainly detected in outer area of fabric layer, where monazite coating is much thicker (~165 nm) compared to inner area (~85 nm) (Figure 11). Why monazite coating thickness of ~85 nm is not enough to provoke fiber-pullout might be explained by the following reasons:

- Porous monazite coating will be partially infiltrated by polysilazane precursor (RTM). The resulting SiCN after pyrolysis will stabilize monazite coating and reduce functionality.
- High fiber roughness of SA3-fibers (~200 nm) requires a minimum monazite coating thickness of >85 nm to effect a weak fiber/matrix-bonding (enables fiber-pullout).

To identify possible Si-C-N enrichments or local formation of lanthanum silicates (ref. Eq. 1), TEM analyses (energy filtered TEM, EFTEM) of fiber/matrix interface before exposure to air have been performed (Figure 12). Mapping images provide a qualitative interpretation of given elements (each image represents one indicated element). Varying sample thickness influences the observed grey levels. Up to a thickness/intensity ratio (t/⊠) of ~0.5 the confidence limit is high. Monazite coating is well bonded to fiber surface. The presence of elements La, P and O indicates the presence of LaPO$_4$. Furthermore, selected area electron diffraction (SAED) TEM analyses (not shown in this study) demonstrate the presence of LaPO$_4$. Nevertheless, due to observed small crystallite size, the local formation of lanthanum silicates cannot be excluded. Next to La, P and O, carbon could be detected in monazite coating which is caused by glue used for sample preparation. This establishes evidence that monazite coating is porous, an observation which is consistent with reported data.[10,16,19] Although there are pores in monazite coating, it is not infiltrated by polysilazane precursor during (re-)infiltration in PIP-process as no Si or N could be observed. Summarizing, due to rough SA3-fiber surface (~200 nm) monazite coating thickness >85 nm is required to obtain weak fiber/matrix bonding which enables fiber-pullout, high strength and damage tolerance.

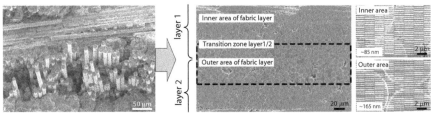

Figure 11. SEM-images of fracture surface as well as microstructure of SA3-F08$_{LaPO4}$/SiCN (5xFC) before exposure to air.

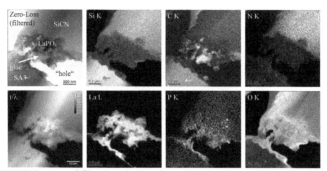

Figure 12. EFTEM-maps of fiber/matrix-interface of SA3-F08$_{LaPO4}$/SiCN (5xFC) before exposure to air.

Before exposure to air, no silica (SiO$_2$) could be observed (by SEM) on SiC-fibers which
might be formed during monazite coating process (ICF) (Figure 13 a). During exposure to air,
there is silica formation on SiC-fibers (~10–120 nm) as well as SiCN-matrix which is exem-
plarily shown for SA3-F08$_{LaPO4}$/SiCN (FC) in Figure 13 b. The presence of monazite coating
inhibits silica growth on SA3-fibers: Silica film is significantly thinner compared to uncoated[20]
or pyc-coated fibers[25] (at least by ~50%). But there is no correlation between monazite coating
thickness and silica thickness on SiC-fibers which is consistent to reported data by Boakye et
al.[16] Therefore, the observed oxidation inhibition of SA3-fibers is unrelated to oxygen diffusion
through monazite but related to contact with phosphoric acid while coating.[20]

Figure 14 shows EFTEM analysis of fiber/matrix interface of SA3-F08$_{LaPO4}$/SiCN (FC)
after exposure to air. Monazite coating is enclosed by silica films on SA3-fiber and SiCN-matrix.
In contrast to unexposed status, monazite coating exhibits no pores (no carbon from glue could
be observed in the coating). Sintering of monazite at high temperatures (exposure temperature is
higher than ICF-temperature (900°C) and maximum pyrolysis temperature (1000°C)) is well
known[14,16,19] and might be supported by silica formation from fibers and matrix. Elements La, P
and O could be detected in the coating indicating for the presence of monazite, which is
consistent with further SAED-TEM analyses (not shown in this study). Locally, there are La-rich
and P-poor spots (indicated by white arrows in La- and P-mappings) located in the outer area of
the coating. This might be caused by local reduction of monazite to La$_3$PO$_7$ (ref. Eq. 3) due to
the presence of free carbon from monazite coating (ref. Figure 2 a), SA3-fiber[29-32] or SiCN-
matrix in consequence of silica formation. Furthermore, locally higher concentrations of Si and
O could be detected (indicated by black arrows in Si- and O-mappings). This can be interpreted
as evidence for the formation of silica or lanthanum silicates in monazite coating.

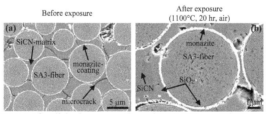

Figure 13. SEM-images of polished cross section of SA3-F08$_{LaPO4}$/SiCN (5xFC) before (a) and
after (b) exposure to air.

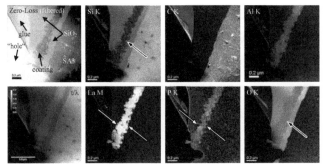

Figure 14. EFTEM-mappings of fiber/matrix-interface of SA3-F08$_{LaPO4}$/SiCN (5xFC) after exposure to air (1100°C, 20 hr).

Compared to unexposed status, aluminum (<1 at.%) is homogeneously distributed in the coating which comes from SA3-fiber (<1 wt.%, added to fibers as a sintering aid[30,32]). The presence of Al could be confirmed with further TEM-EDS analyses (not shown here) and plays a key role for oxidation kinetics.

Mogilevsky et al.[20] proposed an oxidation mechanism for monazite coated SA3-fibers. Due to diffusion of Al and P (monazite is slightly P-rich, ref. Figure 2 a) there is a formation of Al/P-rich layer (e.g. AlPO$_4$ networks, located between silica film of fiber and monazite coating) which acts oxidation inhibiting especially in thin silica films. Oxidation inhibition is not effective anymore if diffusion of P disrupts caused either by limited Al-diffusion or P-depletion of P-source. Due to the fact that neither a P-rich layer (outer border – view from fiber surface – of monazite coating) nor an Al-rich layer (inner border – view from fiber surface – of monazite coating) could be detected in EFTEM-mappings, it seems to be possible that growth of Al/P-rich layer was stopped by P-depletion. With further increasing silica thickness oxidation is predominantly driven by diffusion of oxygen through silica. Finally, oxidation kinetics will be similar to uncoated fibers.[20]

CONCLUSIONS

On the basis of the data reported and discussed, the following conclusions can be drawn:

(1) Due to much better saturation quality of rhabdophane solution in SiC-fabrics, foulard coating process (FC) leads to a much more homogeneous (with respect to covering ratio and gradient in coating thickness) monazite coating compared to dip coating process (DC).

(2) For both monazite coating processes (FC/DC), there is a gradient in coating thickness between filaments located near outer edge of the fabric ("loose" packing) and located in inner bundle ("tight" packing), whereas monazite coating in "loose" packing region is much thicker (at least factor 2) than in "tight" packing region.

(3) To obtain monazite coating thickness (theoretical calculation based on mass gain during coating process) of ~100 nm and 300 nm, 5–7 (for FC) and 14–15 (for DC) coating cycles have to be performed.

(4) SiC$_{LaPO4}$/SiCN based on two different Tyranno SA3 fabric types (SA3-I16 and SA3-F08) were successfully manufactured via PIP-process for the first time.

(5) Homogeneous monazite coating via FC (~100 nm) is related to moderate bending strength (~150–170 MPa) and damage tolerance caused by weak fiber/matrix bonding. Nevertheless, due to high fiber roughness coating thickness has to be thicker than ~85 nm to enable significant fiber-pullout. In contrast to that, inhomogeneous monazite coating via DC (but

much thicker (~300 nm) compared to FC) is related to brittle material behavior with less damage tolerance.

(6) Moderate bending strength (before exposure to air) is related to SA3-fibers strength degradation which is mainly caused by intercoat firings (ICF) during coating process. After 15 ICF, fiber strength is reduced by ~50%.

(7) After exposure to air (1100°C, 20 hr), bending strength of SiC_{LaPO4}/SiCN is reduced by only 16%, which is mainly contributed by oxidation resistant monazite coating.

(8) Microstructural analyses before exposure show that porous monazite is present in the coating. After exposure to air, monazite sintered (no pores present) and is enclosed by silica films from fiber and matrix. Locally, La-rich and P-poor spots are located in the outer area of the coating indicating reduction of monazite probably to La_3PO_7. Furthermore, aluminum (from SA3 fiber) incorporated in the coating which might be an indication for $AlPO_4$ networks decreasing monazite functionality. Furthermore, Al plays a key role for oxidation inhibition especially in thin silica films.

ACKNOWLEDGEMENT

Financial support from the European Commission within the project Aero-Thermo-dynamic Loads on Lightweight Advanced Structures II (ATLLAS II, FP7-263913) is gratefully acknowledged.

REFERENCES

[1] Y. Hikichi, T. Nomura, Melting Temperatures of Monazite and Xenotime, *J. Am. Ceram. Soc.*, **70**, 252–3 (1987)

[2] P.E.D. Morgan, D.B. Marshall, Ceramic Composites of Monazite and Alumina, *J. Am. Ceram. Soc.*, **78**, 1553–63 (1995)

[3] J.B. Davis, D.B. Marshall, R.M. Housley, and P.E.D. Morgan, Machinable Ceramics Containing Rare-Earth Phosphates, *J. Am. Ceram. Soc.*, **81**, 2169–75 (1998)

[4] R.S. Hay, D.B. Marshall, Deformation Twinning in Monazite, *Acta Mater.*, **51**, 5235–54 (2003)

[5] J.B. Davis, R.S. Hay, D.B. Marshall, P.E.D. Morgan, and A. Sayir, Influence of Interfacial Roughness on Fiber Sliding in Oxide Composites with La-Monazite Interphases, *J. Am. Ceram. Soc.*, **86**, 305–16 (2003)

[6] P.-Y. Lee, M. Imai, and T. Yano, Fracture Behavior of Monazite-Coated Alumina Fiber-Reinforced Alumina-Matrix Composites at Elevated Temperature, *J. Ceram. Soc. Jpn.*, **112**, 628–33 (2004)

[7] K.A. Keller, T.-I. Mah, T.A. Parthasarathy, E.E. Boakye, P. Mogilevsky, and M.K. Cinibulk, Effectiveness of Monazite Coatings in Oxide/Oxide Composites after Long-Term Exposure at High Temperature, *J. Am. Ceram. Soc.*, **86**, 325–32 (2003)

[8] T.J. Hwang, M.R. Hendrick, H. Shao, H.G. Hornis, and A.T. Hunt, Combustion Chemical Vapor Deposition (CCVD) of $LaPO_4$ Monazite and Beta-Alumina on Alumina Fibers for Ceramic Matrix Composites, *Mater. Sci. Eng., A*, **244**, 91–6 (1998)

[9] M.K. Cinibulk, Sol-Gel Coating of Nicalon™ Fiber Cloths, *20th Ann. Conf. on Comp., Adv. Ceram., Mater., and Struct.: B: Ceram. Eng. and Sci. Proc.*, John Wiley & Sons, Inc., 241–9 (2008)

[10] E.E. Boakye, R.S. Hay, P. Mogilevsky, and L.M. Douglas, Monazite Coatings on Fibers: II, Coating without Strength Degradation, *J. Am. Ceram. Soc.*, **84**, 2793–801 (2001)

[11] M.K. Cinibulk, Deposition of Oxide Coatings on Fiber Cloths by Electrostatic Attraction, *J. Am. Ceram. Soc.*, **80**, 453–60 (1997)

[12] G.E. Fair, R.S. Hay, and E.E. Boakye, Precipitation Coating of Monazite on Woven Ceramic Fibers: I. Feasibility, *J. Am. Ceram. Soc.*, **90**, 448–55 (2007)

[13]R.S. Hay, E.E. Boakye, and M.D. Petry, Effect of Coating Deposition Temperature on Monazite Coated Fiber, *J. Eur. Ceram. Soc.*, **20**, 589–97 (2000)

[14]R.S. Hay, E.E. Boakye, Monazite Coatings on Fibers: I, Effect of Temperature and Alumina Doping on Coated-Fiber Tensile Strength, *J. Am. Ceram. Soc.*, **84**, 2783–92 (2001)

[15]G.E. Fair, R.S. Hay, and E.E. Boakye, Precipitation Coating of Monazite on Woven Ceramic Fibers: II. Effect of Processing Conditions on Coating Morphology and Strength Retention of Nextel 610 and 720 Fibers, *J. Am. Ceram. Soc.*, **91**, 1508–16 (2008)

[16]E.E. Boakye, P. Mogilevsky, T.A. Parthasarathy, R.S. Hay, J. Welter, and R.J. Kerans, Monazite Coatings on SiC Fibers I: Fiber Strength and Thermal Stability, *J. Am. Ceram. Soc.*, **89**, 3475–80 (2006)

[17]G.E. Fair, R.S. Hay, and E.E. Boakye, Precipitation Coating of Rare-Earth Orthophosphates on Woven Ceramic Fibers - Effect of Rare-Earth Cation on Coating Morphology and Coated Fiber Strength, *J. Am. Ceram. Soc.*, **91**, 2117–23 (2008)

[18]J.B. Davis, D.B. Marshall, and P.E.D. Morgan, Monazite-Containing Oxide/Oxide Composites, *J. Eur. Ceram. Soc.*, **20**, 583–7 (2000)

[19]E.E. Boakye, R.S. Hay, and M.D. Petry, Continuous Coating of Oxide Fiber Tows Using Liquid Precursors: Monazite Coatings on Nextel 720^{TM}, *J. Am. Ceram. Soc.*, **82**, 2321–31 (1999)

[20]P. Mogilevsky, E.E. Boakye, R.S. Hay, J. Welter, and R.J. Kerans, Monazite Coatings on SiC Fibers II: Oxidation Protection, *J. Am. Ceram. Soc.*, **89**, 3481–90 (2006)

[21]M.K. Cinibulk, G.E. Fair, and R.J. Kerans, High-Temperature Stability of Lanthanum Orthophosphate (Monazite) on Silicon Carbide at Low Oxygen Partial Pressures, *J. Am. Ceram. Soc.*, **91**, 2290–7 (2008)

[22]J.R. Mawdsley, J.W. Halloran, The Effect of Residual Carbon on the Phase Stability of $LaPO_4$ at High Temperatures, *J. Eur. Ceram. Soc.*, **21**, 751–7 (2001)

[23]UBE Industries Ltd. (Yamaguchi, Japan), *Tyranno Fiber (Product Catalogue)*, http://www.upilex.jp/e_ceramic.html, Date Accessed 10/12/2012

[24]UBE Industries Ltd. (Yamaguchi, Japan), *UBE Tyranno Grade SA Ceramic High Temperature Fiber*, http://www.matweb.com/search/datasheet.aspx?matguid=77790814f923436e9a327f11a4094c52, Date Accessed 10/12/2012

[25]E. Klatt, A. Frass, M. Frieß, D. Koch, and H. Voggenreiter, Mechanical and Microstructural Characterisation of SiC- and SiBNC-Fibre Reinforced CMCs Manufactured via PIP Method Before and After Exposure to Air, *J. Eur. Ceram. Soc.*, **32**, 3861–74 (2012)

[26]E.J. Opila, R.E. Hann Jr., Paralinear Oxidation of CVD SiC in Water Vapor, *J. Am. Ceram. Soc.*, **80**, 197–205 (1997)

[27]E.J. Opila, Oxidation and Volatilization of Silica Formers in Water Vapor, *J. Am. Ceram. Soc.*, **86**, 1238–48 (2003)

[28]E.J. Opila, Variation of the Oxidation Rate of Silicon Carbide with Water-Vapor Pressure, *J. Am. Ceram. Soc.*, **82**, 625–36 (1999)

[29]T. Ishikawa, Y. Kohtoku, K. Kumagawa, T. Yamamura, and T. Nagasawa, High-Strength Alkali-Resistant Sintered SiC Fibre Stable to 2,200°C, *Nature*, **391**, 773–5 (1998)

[30]T. Ishikawa, S. Kajii, T. Hisayuki, and Y. Kohtoku, New Type of SiC-Sintered Fiber and Its Composite Material, *Ceram. Eng. Sci. Proc.*, **18**, 283–90 (1998)

[31]T. Ishikawa, S. Kajii, T. Hisayuki, K. Matsunaga, T. Hogami, and Y. Kohtoku, New Type of Sintered SiC Fiber and Its Composite Material, *Key Eng. Mater.*, **164–5**, 15–8 (1999)

[32]H. Ichikawa, T. Ishikawa, Silicon Carbide Fibers (Organometallic Fibers), *Compr. Comp. Mater.*, Eds. A. Kelly, C. Zweben, and T. Chou, Elsevier Science Ltd., Oxford, England, **1**, 107–45 (2000)

FIBER, POROSITY AND WEAVE EFFECTS ON PROPERTIES OF CERAMIC MATRIX COMPOSITES

Ojard, G.[1], Cuneo, J.[2]. Smyth, I.[1], Prevost, E.[1], Gowayed, Y.[3], Santhosh, U.[4], and Calomino, A.[5]

[1] Pratt & Whitney, East Hartford, CT
[2] Southern Research Institute, Birmingham, AL
[3] Auburn University, Auburn, AL
[4] Structural Analytics, Inc., Carlsbad, CA
[5] NASA – Langley Research Center, Hampton, VA

ABSTRACT

As insertion opportunities keep increasing for ceramic matrix composites in aerospace and power generation, the understanding of changes in microstructure needs to be explored. Past efforts have looked at the effect of different weave architectures on the mechanical performance of CMCs. This effort expands previous studies by changing the fiber used in the CMC. This change was also accompanied by a process change that affected porosity distribution in the material. The changes introduced were investigated via micro-structural and property characterization. The property characterization consisted of mechanical and thermal testing. The results from this testing will be presented, trends reviewed and analysis done and compared to past testing efforts.

INTRODUCTION

Composites fabricated with a ceramic matrix and ceramic fiber is nearing thirty years in development [1,2] and are gaining wider acceptance for mission critical applications [3]. As applications where Ceramic Matrix Composites (CMCs) are employed, the acceptance of CMCs in industry increases as well as their insertion opportunities. With this increasing interest, the characterization effort has to expand beyond the current point-design approach used for some potential aerospace applications [4-6].

With that goal, work was expanded from testing a range of different weaves [7] from one composite system and changing the fiber while keeping the weave constant (in this case, an 8 harness satin weave at 22 ends per inch). The fiber was changed from CG-Nicalon™ to Hi-Nicalon™. The results of this change and the changes in microstructure and properties will be discussed.

PROCEDURE

Material

With the interest in exploring multiple factors such as changing the fiber, a polymer infiltration pyrolysis system was chosen for the manufacturing. This was due to the ability of the process to be easily transferred to the different CMC systems with limited modifications to the matrix processing that would affect the properties. The material system used for this effort was the SiC/SiNC system with either CG-Nicalon™ to Hi-Nicalon™ fibers (both are non-stoichiometric SiC fibers) with the Hi-Nicalon™ fiber having a higher modulus [8]. The

23

interface coating for both systems had BN present as the weak interface. The matrix of Si, N and C was arrived at by multiple iterations via the polymer pyrolysis process. The initial processing step was resin transfer molding for the CG-Nicalon™ panel while the Hi-Nicalon™ panel was processed using autoclave processing. The panel layup for the Hi-Nicalon™ panel deviated from the standard cross ply layup of [0/90]$_{2s}$ to [0/90]$_4$. The fabrication did not show any panel warpage to indicate that the non-standard ply layup was an issue. Additional material information has been published by the authors [7,9].

Testing – Fiber

Since there was a fiber change between some of the material, fiber testing was done to check the modulus of the fibers. Fiber tows were available and a series of different gage lengths were tested to determine the system compliance as well as the modulus of the fiber in question (ASTM C1557). This effort assumed an average fiber diameter to determine the cross section area (14 μm diameter and 500 fibers per tow for CG-Nicalon™ and Hi-Nicalon™ fibers). The CG-Nicalon™ fibers were tested and reported previously by some of the authors [9].

Testing – Material/Coupon

The testing consisted of a series of mechanical testing: tensile and interlaminar as well as limited thermal testing. All testing was done per ASTM standards. Test temperatures ranged from room temperature to 1200°C. The focus of the analysis in this paper will be on testing at lower temperatures (mainly room temperature). Thermal testing was performed over a range of temperatures.

Characterization:

Characterization for these samples consisted of optical microstructures. Images were taken and image analysis was done to determine porosity. This was done on a series of samples to get representative images and porosity values.

RESULTS

Fiber Testing

The results of the fiber tow testing are shown in Figure 1. As noted earlier, the CG-Nicalon™ fiber testing was done earlier [9] and it is presented here for completeness. As part of the analysis found in ASTM C1557, the intercept determines the system compliance. This testing was done at least seven years apart with the Hi-Nicalon™ fiber being performed the more recently. The test equipments used in the testing of the CG-Nicalon™ and Hi-Nicalon™ fibers were different, and therefore the system compliance in the two cases is different. Past testing showed that the CG-Nicalon™ fiber has an elastic modulus of 190 GPa [9]. The Hi-Nicalon™ fiber done during this testing series was found to have an elastic modulus of 290 GPa. The latter value is in good agreement with the reported literature value of 270 GPa [8].

Material/Coupon Testing

Mechanical Testing

Tensile testing of the CMC with Hi-Nicalon™ fibers (S200-H) was conducted and the average results are shown in Table I. For this series of testing, at least 6 samples were tested at each temperature. The properties start to show a slight decrease as the testing nears 1200°C. In addition, a series of compression tests were also run and the average results of that testing are shown in Table II. For this testing, there were 3 samples tested per temperature. In this series, the strength increases with temperature which was opposite of what was seen in the tensile testing. The trend in the compressive modulus was observed to be similar to that for the tensile modulus.

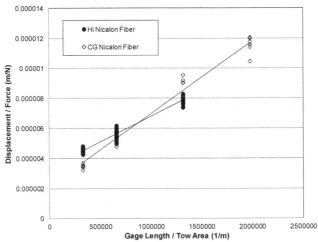

Figure 1. Tow Testing Data of the Hi-Nicalon™ and CG-Nicalon ™ fibers

Table I. Average Tensile Results for Hi-Nicalon™ Tensile Coupons

Test Temp (°C)	Proportional Limit* (MPa)	UTS (MPa)	Modulus (GPa)	Strain to Failure (mm/mm)
21	98.98	528.69	145.6	0.0061
538	96.92	511.37	139.6	0.0066
1204	71.56	469.13	129.0	0.0078

* = determined via 0.005% offset method

Table II. Average Compressive Results for Hi-Nicalon™ Tensile Coupons

Test Temp (°C)	Proportional Limit* (MPa)	Compressive Strength (MPa)	Modulus (GPa)	Strain to Failure* (mm/mm)
21	---	493.42	132.2	---
538	---	496.72	123.3	---
1204	---	596.41	110.9	---

* = data not reported for this series

Interlaminar Tensile Testing

A series of interlaminar tensile tests were done and the results of this testing are shown in Table III. The interlaminar shear testing is not being reported since the correct failure mode was not seen in the testing done on the CG-Nicalon™ CMC, and so no comparison could be done. The results of Table III show that the average interlaminar tensile capability is 9.6 MPa.

Table III. Room Temperature Interlaminar Tensile Results for Hi-Nicalon™ Material

Sample ID	Test Temperature (°C)	Interlaminar Tensile Strength (MPa)
1	21	8.21
2	21	10.86
3	21	9.49
4	21	10.07
5	21	9.76
6	21	9.19
Average		9.60
StDev		0.889

Thermal Testing

In-plane thermal expansion and through thickness thermal conductivity were determined for a select series of samples and the average thermal expansion results are shown in Figure 2. The average through thickness thermal conductivity is shown in Figure 3. The material has low expansion and conductivity as compared to metals and to pure silicon carbide (SiC). This is not unexpected, since the matrix is not pure SiC and there are significant interfaces present (BN around the fibers) as well as porosity in the matrix and tows.

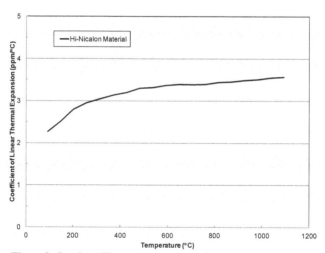

Figure 2. In-plane Thermal Expansion of Hi-Nicalon™ Material

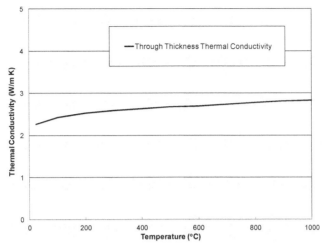

Figure 3. Average Through Thickness Thermal Conductivity of Hi-Nicalon™ Material

Characterization Results

 Typical microstructure images of Hi-Nicalon™ CMC material are shown in Figure 4The images show the presence of porosity in the matrix and within the fiber tows. Image analysis of the material determined the average porosity present in the material as being 4.1%. This is much lower than what has been reported by the authors for the CG-Nicalon™ material where the average porosity was found to be 10% [7].

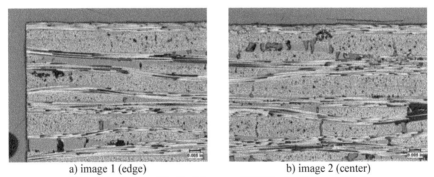

a) image 1 (edge)   b) image 2 (center)
Figure 4. Series of Optical Images of Hi-Nicalon™ Panel (50x)

DATA ANALYSIS/DISCUSSION

Tensile Behavior

As noted previously, the only change between the two materials tested was the fiber; the weave and matrix were the same. A comparison between the room temperature properties of the two CMCs is shown in Table IV. The change to the Hi-Nicalon™ fiber is observed to have increased all the properties recorded. What is not clear is the effect of porosity since it was found that the Hi-Nicalon™ material had an average porosity of 4.1% while the CG-Nicalon™ material had an average porosity of 10% [7]. Based on the past work where the constituent properties were known, the elastic modulus was modeled using a rule of mixtures approach [7]. That work can be extended here since the fiber properties were confirmed via testing. The results of this are shown in Figure 5. As was the case of the past work of the authors [7], the values seen in testing can be predicted using a rule of mixtures calculation. Additional analysis is needed on the other properties but it is clear that the change in fiber resulted in an increased capability material. The compressive properties are in close agreement with the tensile data documented and modeled by rule of mixtures.

Table IV. Average RT Tensile Results for CG-Nicalon™ and Hi-Nicalon™ Coupons

Fiber	Proportional Limit* (MPa)	UTS (MPa)	Modulus (GPa)	Strain to Failure (mm/mm)
CG-Nicalon™	88.7	268	115.4	0.0045
Hi-Nicalon™	98.98	528.69	145.6	0.0061

* = determined via 0.005% offset method

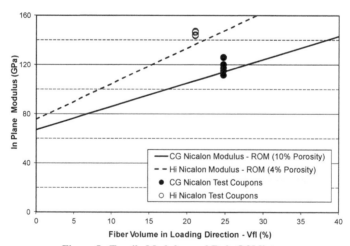

Figure 5. Tensile Modulus and Rule Of Mixtures

Interlaminar Tensile Behavior

The interlaminar tensile behavior of the Hi-Nicalon™ is shown in Table III and the average strength is 9.6 MPa. This differs from the previous testing by the authors on CG-Nicalon™ where the average value was 14.1 MPa. It is not clear that this difference is significant or not considering the wide range of values that this testing can generate [10]. This is an area where additional testing and test method development would be beneficial.

Thermal Behavior

A comparison of the in-plane coefficient of thermal expansion (α) between the two material sets is shown in Figure 6. The data does not differentiate between the two CMCs except at the lower temperatures. The only difference in the two material systems studied here is the porosity and fiber type. In-plane properties, are expected to be fiber dominated and there is not a significant difference between the fibers (α of 3.2 ppm/°C for GC-Nicalon™ and α of 3.5 ppm/°C for Hi-Nicalon™ at 500°C) [8,11].

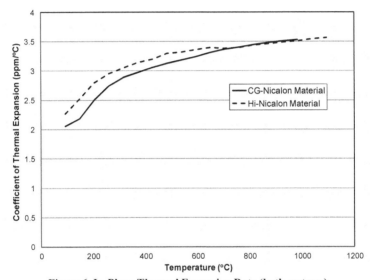

Figure 6. In-Plane Thermal Expansion Data (both systems)

A comparison of the through thickness thermal conductivity is shown in Figure 7. Here there is a clear distinction between the two materials systems as this test direction is dependent both on the fibers and on the porosity. The porosity has a slight change but the conductivity of the fibers is clearly different as Hi-Nicalon™ has 5 times the conductivity of the CG-Nicalon™ fiber so the difference here is in line with the increase due to the fiber and some increase due to lower porosity.

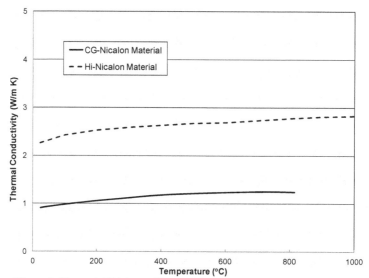

Figure 7. Through Thickness Thermal Conductivity Data (both systems)

CONCLUSION

A series of tests were conducted on 8HS panels fabricated with either CG-Nicalon™ or Hi-Nicalon™ fibers. The change to higher modulus fibers resulted in higher in-plane tensile properties (some of which could be modeled used on rule of mixtures expressions) for the composite, but there was a decrease in corresponding interlaminar properties. Some of the differences could be also due to the difference in porosities between panels from the two CMC material systems. Thermal testing showed that there was not much change in the in-plane thermal expansion but the higher through-thickness thermal conductivity seen in the S200-H testing was due to the higher conductivity of the Hi-Nicalon™ fiber.

This work shows the benefit of looking at different fibers, which can impact the design space for future applications.

FUTURE WORK

Additional modeling and durability testing should be done to look at further refinement of the opening of the design space.

ACKNOWLEDGMENTS

Work performed under the Ceramic Matrix Composite Turbine Blade Demonstration Program, Contract FA8650-05-D-5806 Task 0021, Dr. Paul Jero program manager.

REFERENCES

1. Richerson, D.W., Iseki, T., Soudarev, A.V. and van Roode, M., "An Overview of Ceramic Materials development and Other Supporting Technologies", Chapter 1 from Ceramic Gas Turbine Component Development and Characterization, van Roode, M., Ferber. M.K. and Richerson, D.W., Eds. ASME Press, New York, 2003

2. Landini, D.J., Fareed, A.S., Wang, H., Craig, P.A. and Hemstad, S., "Ceramic Matrix Composites Development at GE Power Systems Composites, LLC", Chapter 14 from Ceramic Gas Turbine Component Development and Characterization, van Roode, M., Ferber. M.K. and Richerson, D.W., Eds. ASME Press, New York, 2003

3. Kestler, R. and Purdy, M., "SiC/C for Aircraft Exhaust", presented at ASM International's 14th Advanced Aerospace Materials and Processes Conference, Dayton, OH, 2003

4. Brewer, D., 1999, "HSR/EPM Combustion Materials Development Program", Materials Science & Engineering, 261(1-2), pp. 284-291.

5. Brewer, D., Ojard, G. and Gibler, M., "Ceramic Matrix Composite Combustor Liner Rig Test:, ASME Turbo Expo 2000, Munich, Germany, May 8-11, 2000, ASME Paper 2000-GT-670

6. Verrilli, M. and Ojard, G., "Evaluation of Post-Exposure Properties of SiC/SiC Combustor Liners testing in the RQL Sector Rig", Ceramic Engineering and Science Proceedings, Volume 23, Issue 3, 2002, p. 551-562.

7. Ojard, G., Prevost, E., Santhosh, U., Naik, R., and Jarmon, D. C., "Weave And Fiber Volume Effects In A PIP CMC Material System", in Mechanical Properties and Performance of Engineering Ceramics and Composites VII (eds D. Singh, J. Salem, M. Halbig and S. Mathur), John Wiley & Sons, Inc., Hoboken, NJ, USA. doi: 10.1002/9781118217467.ch20

8. http://www.coiceramics.com/sicfibers.html

9. Ojard. G., Rugg, K., Riester, L., Gowayed, Y. and Colby, M., "Constituent Properties Determination and Model Verification for a Ceramic Matrix Composite Systems", Ceramic Engineering and Science Proceedings. Vol. 26, no. 2, pp. 343-350. 2005.

10. Ojard. G., Unpublished Research, Pratt & Whitney, East Hartford, CT.

11. http://www.fusion.ucla.edu/apex/pdfs/SiC.pdf

WEAVE AND FIBER VOLUME EFFECTS ON DURABILITY OF CERAMIC MATRIX COMPOSITES

Ojard, G.[1], Prevost, E.[1], Santhosh, U.[2], Naik, R.[1], and Jarmon, D. C.[3]

[1] Pratt & Whitney, East Hartford, CT
[2] Structural Analytics, Inc., Carlsbad, CA
[3] United Technologies Research Center, East Hartford, CT

ABSTRACT

With the increasing interest in ceramic matrix composites for a wide range of applications, fundamental research is needed in the areas of multiple weaves and fiber volume. Understanding how the material performs in durability with differing weaves and fiber volume will affect how the material is viewed for any insertion application. With this in mind, a series of three panels were fabricated via a polymer infiltration process: 8 harness satin (HS) balanced symmetric layup, 8 HS bias weave, and angle interlock. From these panels, a series of durability tests were undertaken with the sample oriented in both the warp and fill direction. These tests consisted of creep and 30 Hz fatigue. The results from this testing will be presented, trends reviewed and analysis done.

INTRODUCTION

As interest grows for increasing temperature capable materials such as Ceramic Matrix Composites (CMCs), the need to extend the testing and to understand the material effects increases [1]. This is especially true where there are designs or applications that can take advantage of the low density and/or high temperature capability of the material [2,3]. Past work of the authors looked at weaves and fiber volume effects on select mechanical properties [1]. The tensile testing efforts of that material showed that the modulus agreed with the rule of mixtures equation and that the proportional limit and ultimate tensile strength increased with fiber volume. This expands the design space from a point design approach.

As a follow on to that effort, durability testing (creep and fatigue) was done at elevated temperatures to further explore the weave and fiber volume effects that were seen in the fast fracture results [1]. The panels were the same: cross ply balanced panel, bias panel with a ratio of 3:1 and an angle interlock panel. The results of the creep and fatigue testing are reported and conclusions drawn.

PROCEDURE

Material

For this testing series, there were three panels that were fabricated and machined into tensile coupons for durability testing: cross ply balanced panel, bias panel with a ratio of 3:1 and an angle interlock panel [1]. The material system used for this effort was the SiC/SiNC composite comprising a non-stochiometric SiC (CG Nicalon™) fiber, coated with BN, in a matrix of Si, N and C manufactured by multiple iterations of a polymer pyrolysis process. This material has been characterized previously by some of the authors [4]. The baseline panel for this effort was a cross ply panel using a 22 ends per inch (epi) 8 HS balanced cloth. The panel was a 6 ply panel with an overall fiber volume set at 40%. The second panel was a Bias weave panel (cross ply layup at 6 plys) where the warp fibers were set at ~3x the fill direction fibers. The overall fiber volume was set at 40% (consistent with the baseline panel). The Angle Interlock panel was an effort to combine a bias weave with a low angle interlock to increase interlaminar properties. The total fiber volume was set at 35%. The manufacturing goals for the fiber volume which shows the nature of the various weaves is found in Table I.

Fiber Volume (Past Characterization):
    As noted earlier and in the previous paper [1], the bias and angle interlock panels were fabricated with fiber volumes that were different in the warp and fill direction. The fiber volume effort was determined by analysis of the stress-strain curves [1,5] and image analysis [1]. The results of this testing are shown in Table I with the expected values set during manufacturing. For the analysis to be reported in the remainder of the paper, the fiber volume based on the tensile curves will be used based on the good agreement with the optical work that was not extended to the other panel types (See Table I.). In addition, a porosity analysis was done on the samples and no differentiation was seen between the panel types with the overall average being 9.9%. The key mechanical property is also shown in Table I: proportional limit [1].

Table I. Manufacturing Goals, Fiber Volume and Proportional Limit for Panel Types

Panel Type	Fiber Direction	Vf Manufacture (%)	Vf Tensile (%)	Vf Optical (%)	Proportional Limit (MPa)
Baseline	Warp	20.1	24.7	24.7	81.4
	Fill	20.1	24.7	24.7	90.9
Bias	Warp	28.8	35.1	na	105.0
	Fill	11.1	6.9	na	45.7
Angle Interlock	Warp	24.5	21.0	na	89.6
	Fill	10.5	8.4	na	55.6

na = not measured

Durability Testing
    The durability testing consisted of creep testing and fatigue testing. The creep testing was done at 649°C and 982°C. If the samples achieved 100 hours, the testing was stopped. The fatigue testing was done at 30 Hz with a R ratio of 0.05. All fatigue testing was done at 649°C. If the tests hit 2.5 million cycles, the testing was stopped. The test matrix for this effort is shown in Table II. All testing was done to ASTM standards. There were two stress levels for each test type, specimen orientation and temperature. Some of the repeats were hindered due to lack of material. The bias panel had a delamination that eliminated testing in the warp direction and it also had some impact on the fill direction testing. Every effort was taken to execute the test matrix completely.
    During testing, strain was recorded using extensometers. This was even done during the fatigue testing by decreasing the frequency to 1 Hz in order to allow loop collection on a periodic basis (at decade intervals).

Table II. Durability Test Matrix for the Three Panel Types

Test Description	Test Direction	Temp (°C)	Stress Levels*	Test Environment	Test Method
Creep	Warp	649	2	air	C1337
Creep	Fill	649	2	air	C1337
Creep	Warp	982	2	air	C1337
Creep	Fill	982	2	air	C1337
Fatigue (30 Hz)	Warp	649	3	air	C1360
Fatigue (30 Hz)	Fill	649	3	air	C1360

* = effort was for two tests at each stress level

RESULTS

Creep Testing
    As noted in the test matrix, creep testing was done on samples from all three panels. The baseline testing was done at stresses of 104 and 124 MPa at temperatures of 649°C and 982°C. All the samples tested at 104 MPa achieved the 100 hour test duration (run-out). There were a few failures at the stress level of 124 MP and they are shown in Figure 1 (the predominance of failures occurred at 982°C).
    The testing on the Bias panel was not as complete as that for the baseline panel due to several samples being lost to a delamination in the panel. As noted earlier, this affected the testing only in the warp direction as samples were not available for testing. This was not considered an issue since this direction was expected to be very similar to the fiber loading to the baseline panel. (As can be seen in Table II, this was not the case as the warp direction had a higher fiber volume than expected but this was not known until the testing was nearing completion.) The main focus was therefore on testing in the fill direction that had lower fiber volume. The fill direction testing was done at stresses of 52 and62 MPa at temperatures of 649°C and 982°C. All of the failures occurred at the 62 MPa stress as shown in Figure 2.
    The testing of the Angle Interlock samples generated more failures than was seen for the Baseline and Bias panels. For the warp direction, testing was done at 104 and 124 MPa at temperatures of 649°C and 982°C. The failures were predominantly at the stress level if 124 MPa. For the fill direction, the test temperatures were the same but the stress levels were 52 and 62 MPa consistent with the Bias testing effort. For the fill direction, all but two samples failed during the testing. The results of this effort are shown in Figure 3.

Fatigue Testing
    The fatigue testing was limited in that there was only testing done at one temperature (649°C). The results of the fatigue testing of the baseline panel are simple in that a failure was seen at 145 MPa. The other tests at 104 and 124 MPa resulted in a run-out condition. (The tests were stopped at 2.5 million cycles.) This is shown in Figure 1.
    The Bias testing had some warp direction testing done at stresses of 104 and 124 MPa. All of those tests resulted in sample run-out. Testing in the fill direction resulted in failures except at the lowest stress level of 52 MPa. The results of this testing are shown in Figure 2.

Consistent with the creep testing of the Angle Interlock panel, the fatigue testing generated more failures than seen in the other two panel types. Run-outs were only seen in the warp direction testing at 104 MPa and the fill testing had run-outs at 52 and 62 MPa. The results of this testing are shown in Figure 3.

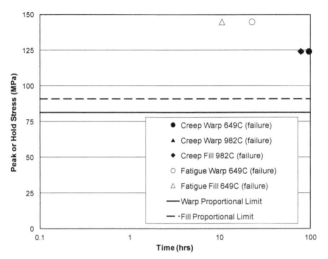

Figure 1. Baseline Panel Durability Data (Failed Data Points)

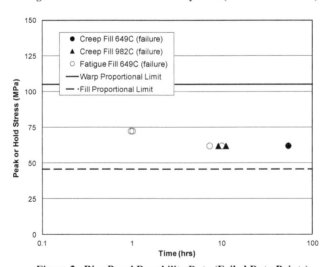

Figure 2. Bias Panel Durability Data (Failed Data Points)

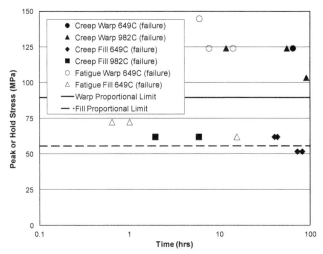

Figure 3. Angle Interlock Panel Durability Data (Failed Data Points)

DATA ANALYSIS/DISCUSSION

In considering the breadth of data presented in conjunction with the differing fiber volumes, the starting point is to look at the data against the appropriate proportional limit for the testing direction. A review of the proportional limit data from past effort showed that the proportional limit was clearly not dependent on temperature allowing the full data to be averaged regardless of temperature [1]. These results are found in Figures 1-3 with the proportional limit clearly marked on the Figures. As noted previously, the samples from the Baseline panel had the fewest failures and these only occurred at the higher stress. This can be seen clearly in Figure 1 (in conjunction with Figures 2 and 3). For the Bias panel samples, more failures were present (fill direction only since the warp direction samples were lost to delamination) and the failures occurred at stress levels closer to the proportional limit than seen in the Baseline testing. In addition, failures occurred much earlier in time. This trend was seen to a greater extent in the Angle Interlock samples (See Figure 3). Failures were more prevalent and occurred at even lower stress levels than for the other two cases. In some specimens, failure even occurred below the proportional limit. These trends are significant since the testing was very short term in nature: 24 hours for fatigue testing and 100 hours for creep testing.

To gain additional insight into the testing, the run-out samples from the creep series were reviewed non-destructively via thermography [6]. This was done to see if there was some indication of damage present in the samples especially for the Bias and Angle Interlock coupons. The thermography color scale was set so that blue corresponded to areas of high diffusivity, indicating the relative absence of defects, and red corresponded to regions of low diffusivity, indicating the presence of defects. What was surprising in the Baseline samples (Figure 4) was that there were more indications for the testing done at 649°C than seen at the 982°C data. (Some of the images appeared to be bowed and this is an artifact of the data collection and not any dimensional changes seen in the sample.) In looking at the Bias samples, some of the testing at 982°C did show some indications but they were in the transition region and not the gauge section of the sample (Figure 5). The Bias samples at 649°C showed changes but not to the extent seen in the Baseline effort. The Angle Interlock showed even less of an effect (Figure 6).

In order to understand why more damage was seen in the thermography images at 649°C than at 982°C, the strain time data for the effort was reviewed. For the Baseline effort typical creep data for the two temperatures are shown in Figure 7. It is clear that more strain is being evolved at the lower temperature (649°C). All the run-out curves were reviewed and the average total strain evolved during the testing for the three panel sets tested is shown in Table III. For the Baseline panel testing, the strain evolved at 649°C is higher than equivalent testing done at 982°C. The limited Bias testing in the warp direction is in line with this conclusion. Testing of the lower fiber volume directions in the Bias and Angle Interlock panels are not as clear cut. This is also the case for the Angle Interlock Warp direction testing, which has higher strains.

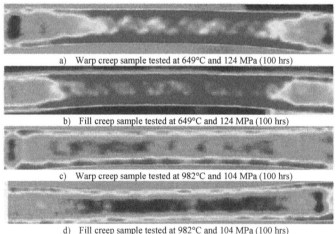

a) Warp creep sample tested at 649°C and 124 MPa (100 hrs)

b) Fill creep sample tested at 649°C and 124 MPa (100 hrs)

c) Warp creep sample tested at 982°C and 104 MPa (100 hrs)

d) Fill creep sample tested at 982°C and 104 MPa (100 hrs)

Figure 4. Baseline Creep Thermography Images Post Run-Out

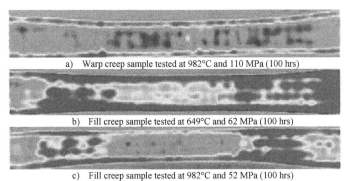

a)   Warp creep sample tested at 982°C and 110 MPa (100 hrs)

b)   Fill creep sample tested at 649°C and 62 MPa (100 hrs)

c)   Fill creep sample tested at 982°C and 52 MPa (100 hrs)

Figure 5. Bias Creep Thermography Images Post Run-Out

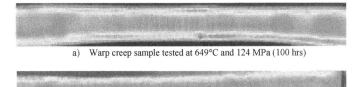

a)   Warp creep sample tested at 649°C and 124 MPa (100 hrs)

b)   Fill creep sample tested at 982°C and 62 MPa (100 hrs)

Figure 6. Angle Interlock Creep Thermography Images Post Run-Out

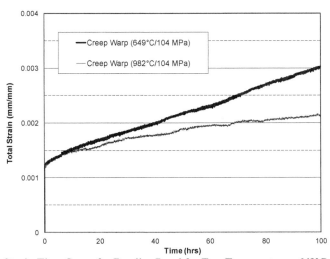

Figure 7. Strain-Time Curve for Baseline Panel for Two Temperatures: 649°C and 982°C

Table III. Evolved Strain during Creep Testing

Panel	Test	Test	Hold	Total
	Direction	Temperature	Stress	Evolved Strain
		(°C)	(MPa)	(mm/mm)
Baseline	Warp	649	104	0.0037
Baseline	Warp	649	124	0.0039
Baseline	Warp	982	104	0.0018
Baseline	Fill	649	104	0.0029
Baseline	Fill	649	124	0.0034
Baseline	Fill	982	104	0.0019
Baseline	Fill	982	124	0.0037
Bias	Warp	982	110	0.0018
Bias	Fill	649	52	0.0030
Bias	Fill	649	62	0.0038
Bias	Fill	982	52	0.0021
Angle	Warp	649	104	0.0023
Angle	Warp	649	124	0.0031
Angle	Warp	982	104	0.0023
Angle	Fill	982	52	0.0034

It is clear that the testing done at 649°C is being influenced by the mechanism of Intermediate Temperature Oxidation (ITO) [7]. The evolved strain at this temperature is not in line with perceived expectations of testing at such a low temperature for a CMC material. Other investigators have looked at this material from creep tests and proposed an ITO issue for testing done at 815°C [8]. That work did not show a relation between the strain evolved during testing and the residual strain unlike work by other investigators who had a quantified relationship in a material system that does not show ITO attack [9]. The present effort is the first time that such an effect has been seen via a non-destructive method. The diffusivity change seen in the current study may be indicative of oxidation attack of the matrix. (Preliminary X-ray Computed Tomography (CT) scanning of the samples is not showing differentiation at this time and more focused CT is being considered.)

What is not clear is the differentiation between the weaves on this effect and on durability. For the Bias and Angle Interlock material, the differentiation in the strain-time curves is not as drastic as for the Baseline material. As seen in Figure 5, the Bias material shows less uniform attack with some of the attack being seen outside of the gauge section. (These areas are also show up in the through transmitted thermography so it is not a surface effect.) The Angle Interlock material shows even less attack even though the evolved strain was similar to the other materials. This may be due to micro-structural differences between the two types of weaves. Optical images looking in the fill direction have been reported previously from the authors and are shown in Figure 8 (for completeness) [1]. It is clear from the images and tensile analysis that there are reduced fiber tows (reduced Vf) in the fill direction. The lack of a pipeline diffusion path for oxidation ingress would explain the differences in oxidation. Additionally, if the matrix was being oxidized uniformly along the edges at these low temperatures, it could hinder attack in the center of the material. Optical cross section work should be done to confirm this.

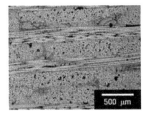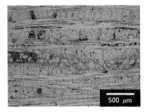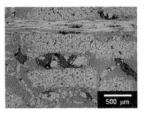

a) Baseline (50x)   b) Bias (50x)   c) Angle Interlock (50x)

Figure 8. Optical Images of Fill Direction for the Three Panel Types
(Image from Reference 1)

Overall for the creep testing done, the baseline panels performed the best during the short duration of the tests. The samples were tested well above the proportional limit with the vast majority of the samples reaching the run-out condition of 100 hours. Residual testing on these samples could reinforce this conclusion.

The fatigue testing due to its nature was much shorter in duration with no test exceeding 24 hours. Therefore, time dependent effects were not expected in the thermography images. Some select images from the fatigue testing are shown in Figure 9 for the three panel types. These images do not have the same appearance as the creep samples that were tested (Figures 4-6). There appears to be some sort of crack or other defect being generated during fatigue testing of the specimens. This is seen with some line indications in Figure 9a and 9b.

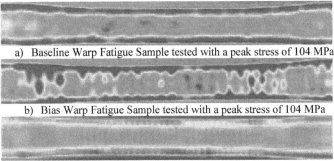

a) Baseline Warp Fatigue Sample tested with a peak stress of 104 MPa

b) Bias Warp Fatigue Sample tested with a peak stress of 104 MPa

c) Angle Interlock Warp Sample tested with a peak stress of 104 MPa

Figure 9. Thermography Images from Fatigue Samples

Based on this, the strain data measured during the fatigue testing was reviewed to see if damage was occurring in the material. A secant analysis was performed. An example of this is shown in Figure 10 for the Baseline sample. The sample shows an initial decrease in the secant modulus during the initial loading followed by a region where the secant is relatively constant. This is followed by a region near the end of testing where the modulus starts to decrease again.

This was seen to occur near 400,000 cycles. (The data spikes are due to the equipment being changed over to 1 Hz data collection to garner hysteresis loops.)

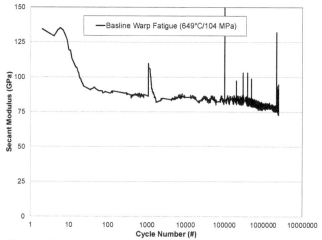

Figure 10. Secant Modulus Analysis of Baseline Fatigue Sample
(run-out sample)

This analysis was also being done for the fill direction, which has a lower fiber volume fraction. The NDE for the sample is shown in Figure 11 and the secant modulus analysis is shown in Figure 12. The figures show more significant degradation occurring in the fill direction specimens towards the end of the test. This observation is consistent with the fact that there were few run-outs achieved at the lower fiber volumes.

a) Bias Fill Sample tested with a peak stress of 52 MPa
Figure 11. Thermography Image from Fatigue Sample

The fatigue testing indicates that some sort of matrix cracking or fracture of fibers is occurring leading to a decrease in the stiffness. The presence of cracks or fiber fractures has not been confirmed at this time.

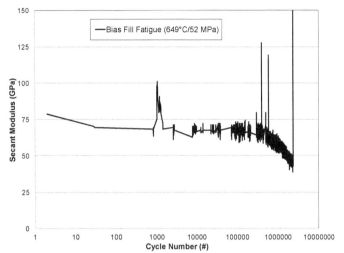

Figure 12. Secant Modulus Analysis of Baseline Fatigue Sample
(run-out sample)

CONCLUSION

A series of durability tests were conducted on an 8HS Baseline panel, a Bias panel and an Angle Interlock panel. Testing was done in both the warp and fill directions. The durability testing showed that at exposure times greater than 24 hours an Intermediate Temperature Oxidation was occurring. This was clearly seen in the Baseline samples. ITO was seen to a lesser extent in the Bias and Angle Interlock samples and this is postulated to be due to the reduced fibers and tows in the loading direction in these specimens that limits pipeline diffusion. The fatigue testing showed that cracking was occurring in the material as indicated by NDE as well as a decrease in the secant modulus at high cycle times.

The overall best durability was seen for the Baseline panel for the times tested. The Bias and Angle Interlock panels showed that they were more susceptible to failure even for testing near the proportional limit.

FUTURE WORK

Future testing will focus on residual property determination and micro-structural examination of internal oxidation.

ACKNOWLEDGMENTS

Initial work performed under the Ceramic Matrix Composite Turbine Blade Demonstration Program, Contract FA8650-05-D-5806 Task 0021, Dr. Paul Jero program manager. Additional work and data analysis was done under Contract FA8650-12-C-5108, Dr. George Jefferson program manager. The authors are indebted to Trish Smyth of Pratt & Whitney and Yasser Gowayed of Auburn University for their insights into the paper during the review process.

REFERENCES

1. Ojard, G., Prevost, E., Santhosh, U., Naik, R., and Jarmon, D. C., "Weave And Fiber Volume Effects In A PIP CMC Material System", accepted into Ceramic Engineering and Science Proceedings, 2012.
2. K.K. Chawla (1998), Composite Materials: Science and Engineering, 2nd Ed., Springer, New York
3. Ashby, Michael F. (2005). Materials Selection in Mechanical Design (3rd Edition). pp: 40. Elsevier, Boston
4. Ojard. G., Rugg, K., Riester, L., Gowayed, Y. and Colby, M., "Constituent Properties Determination and Model Verification for a Ceramic Matrix Composite Systems", Ceramic Engineering and Science Proceedings. Vol. 26, no. 2, pp. 343-350. 2005.
5. Ojard, G., Gowayed, Y., Morscher, G., Santhosh, U., Ahmad, J., Miller, R. and John, R., "Creep and Fatigue Behavior of MI SiC/SiC Composites at Temperature", Published in Ceramic Engineering and Science Proceedings, 2009.
6. Sun, J.G., Stephan, R.R., Verrilli, M.J., Barnett, T. and Ojard, G.C., "Nondestructive Evaluation of Ceramic Matrix Composite Combustor Components", 29th Annual Review of Progress in Quantitative NDE, Volume 22B, Bellingham, WA, July, 14-19, 2002, Melville, NY, American Institute of Physics, 2003, p. 1011-1018
7. Sun, E. Y., Lin, H., & Brennan, J. J. (1997). Intermediate-temperature environmental effects on boron nitride-coated silicon carbide-fibre-reinforced glass-ceramic composites. J.Am.Ceram.Soc.Vol.80, no.3, pp.609-614.1997, 80(3), 609-614
8. Sinnamon, K., Ojard, G., Flandermeyer, B. and Miller, R., "Intermediate Temperature Refinement", published in 2009 Materials Science and Technology Conference and Exposition (MS&T 2009).
9. Ojard, G., Calomino, A., Morscher, G., Gowayed, Y., Santhosh, U., Ahmad J., Miller, R. and John, R., "Post Creep/Dwell Fatigue Testing of MI SiC/SiC Composites", Ceramic Engineering and Science Proceedings, pp. 135-143. 2008

COOLING PERFORMANCE TESTS OF A CMC NOZZLE WITH ANNULAR SECTOR CASCADE RIG

Kozo Nita[1], Yoji Okita[1], Chiyuki Nakamata[1]
Advanced Technology Department, Research & Engineering Division, Aero-Engine & Space Operations, IHI Corporation
229, Tonogaya, Mizuho-machi, Nishitama-gun, Tokyo, 190-1297, JAPAN

ABSTRACT

Ceramic Matrix Composites (CMC) are a promising material because the allowable temperature is 200°C higher than that of conventional Ni-based super-alloys. Many attempts to apply this ceramic to hot section components of advanced gas turbine engines have been made. However, in order to fully gain the performance benefits from this advanced material, more understanding and advancement is needed. One item is to confirm the cooling performance of cooled CMC components like turbine nozzles. For conventional cooled nozzles, internal and external cooling is used. This paper presents two kinds of cooling performance tests with CMC turbine nozzles: one with only impingement cooling and the other with both impingement and film cooling.

First, the impingement cooling test of CMC nozzle was conducted in the steady hot mainflow, with room temperature coolant in order to investigate the internal cooling performance of the nozzle. The impingement cooling Reynolds number was set at 1×10^4, which is consistent with that of actual gas turbine engines. The exit Mach number of the cascade was set at 1 approximately.

Second, the test with both impingement and film cooling of the CMC nozzle was conducted. The film cooling was designed with its impingement cooling Reynolds number being equal to that in the impingement cooling performance test.

The experimental results are presented and discussed in detail in this paper.

INTRODUCTION

The Turbine inlet gas temperature (TIT) has been increasing higher and higher to improve the efficiency of gas turbine engines. Even with this TIT increase, cooling air consumption should be kept at a minimum to achieve higher gas turbine efficiency. Therefore, high temperature gas turbines have been developed with progress in advanced cooling and material technologies. Ceramic Matrix Composite (CMC), such as SiC/SiC, can be applied at much higher temperature and have much lower density as compared to conventional nickel-based super-alloys. So, it is one of the appropriate candidates for hot section components in the next generation gas turbines / aero-engines. Recently, the capability of CMC as a material for turbine gas path components has been reported in several research reports. F. Zhong and G.L. Brown [1] tested the cooling effectiveness of a multi-hole cooled CMC plate. The durability of CMC vanes were demonstrated in some burst tests [2] or thermal-cycle tests [3, 4].

The authors' group has also investigated the CMC's capability [5, 6] in several research and development programs to realize CMC turbine components. The cooled CMC vane research presented in this paper has been carried out under a contracted research program with the Ministry of Economy, Trade and Industry (METI) from 2009. In 2010, the cooled CMC nozzle was designed and manufactured. The vane was tested in steady hot gas flow to evaluate the cooling performance. One of the primary objectives of this cooling performance test is to clarify the effect of unique surface feature of the CMC. In the internal cooling test, only the impingement cooling is applied (Nita et. al. [7]). Impingement cooling can achieve a relatively high heat transfer rate without any complicated internal structure, which is favourable for the CMC nozzles.

Thermal conductivity of CMCs is much lower than that of super-alloys. Therefore, to enhance the cooling performance of the CMC, internal cooling should be combined with external cooling (film cooling). So, film cooling performance with CMC components should also be investigated in order to realize their practical use.

A few studies of the CMC cooling structure have been reported, and they show only basic examination with simple shapes such as the flat-plate shown in reference [1].

In contrast, this is the first report to discuss cooling performance in detail under the conditions which simulates an actual engine, with an actual engine nozzle's shape.

This paper presents the experimental results with the impingement cooling only and with both the impingement and film cooling of the CMC turbine nozzle.

Design of CMC turbine nozzle

In studying the applicability of CMC to the turbine components, a nozzle vane was selected for the present research. Nozzle vanes are exposed at higher temperature than other parts, and thus the cooling performance is important. With all the available internal cooling methods (such as PIN-FIN cooling, Rib-turbulated cooling etc.) being investigated, impingement cooling with a separate metal insert seemed to most suit the CMC nozzle because of its high cooling effectiveness and its relatively simple internal geometry of the vane shell. Any complicated internal structure is not feasible with CMC components.

The external appearance of the designed CMC nozzle is shown in Figure 1. Also, the internal structure of the CMC nozzle with an insert is shown in Figure 2.

The arrangement and the geometry of the impingement holes are designed so that the impingement Reynolds number is consistent with that of typical engines. The geometrical parameters of the impingement holes are same in all the tests. The impingement holes locate on the insert as shown in Table 1, and the insert itself locates from -83%CaX to 83%CaX (Note that negative value corresponds to the PS and positive value to the SS.).

The film cooling holes are designed as the impingement Reynolds number is equal to that of the case with only the impingement cooling, by varying the pressure of the supplied coolant. Film cooling holes are designed as shown in table 2.

Table1 Parameter of impingement hole

	No.1	No.2	No.3	No.4	No.5	No.6	No.7	No.8	No.9
%CaX	−54	−45	−10	0	17	36	52	63	70
Geometry	circle	circle	circle	circle	circle	circle	circle	circle	circle

Table2 Parameter of film hole

	No.1	No.2	No.3	No.4	No.5	No.6	No.7
%CaX	−72	−65	−56	−41	25	40	60
Geometry	circle	circle	circle	circle	circle	circle	circle

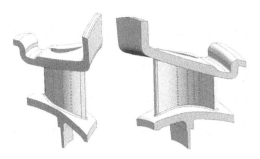

Figure 1 Structure of the CMC nozzle   Figure 2 CMC nozzle with the insert

Test facility

Figure 3 shows the layout of the test facility. The facility for the cooling performance tests can vary the mainflow temperature from about 200°C to 500°C. In the two cooling performance tests, to achieve uniform temperature and pressure distribution, the mainflow, heated by a combustor, was supplied through a mixer to flatten the distribution. Downstream of the mixer, there is a bypass line to broaden control the range of the main gas flow rate and pressure in the test section.

Figure 4 depicts the cross section of the test rig and Figure 5 shows the test rig installed in the facility. The main gas through the mixer flows into the flow straighter and contracts to the passage of the test section. Downstream of the test section, the main gas flows through a diffuser section and discharges outside.

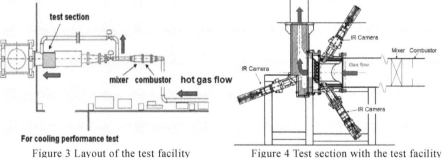

Figure 3 Layout of the test facility   Figure 4 Test section with the test facility
in the thermal shock test

Figure.5 Cascade of the turbine nozzles

Figure 6 shows the PTF (pattern factor), which is an index of uniformity of the gas temperature distribution. To measure the distribution of the radial and circumferential direction, three combination probes were installed in the test section. PTF is defined as,

$$PTF = \frac{T_{g,\max} - T_{g,ave}}{T_{g,ave} - T_{c,in}} \qquad (1)$$

There was only a little temperature non-uniformity; thus, local gas temperatures over the surface of the nozzles were assumed to be the same as the inlet average temperature (T_g).
Cooling effectiveness is defined as,

$$\eta = \frac{T_g - T_{w,g}}{T_g - T_{c,in}} \qquad (2)$$

Because of the uniformity in thermal field, evaluation of cooling effectiveness was possible in only two-dimensional calculation at the mean cross-section.

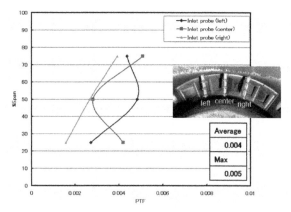

Figure 6   PTF in the cooling performance test

Cascade test rig

As shown in Figure 7, in the test section, the cascade had five nozzles and two side walls. A CMC turbine nozzle was placed in the center of the cascade and the others were dummy metal nozzles.

Three total temperature and pressure combination probes were installed at the inlet of the cascade section. Each probe had three temperature and pressure sensors arranged in the radial direction (nine temperature and pressure sensors in total). Furthermore, three IR cameras set on the sight tubes (observation windows) were used to measure the thermal fields of the pressure and the suction surfaces of the CMC nozzle. Figure 8 shows the arrangement of the three IR cameras. In this paper, the symbols (①~③) correspond to the viewpoints shown in the figure. Blackbody paint with emissivity of 0.94 was applied to the surfaces of the turbine nozzles to capture accurate temperature. Additionally, three thermocouples were attached on the CMC nozzle surfaces to calibrate the temperature fields captured by the IR cameras. Another thermocouple was attached on the pressure side of the trailing edge where IR camera was not able to cover. Small bumps were produced at the thermocouples wiring. There were two markings on each PS and SS to identify the location in the IR camera view. The stays of the IR cameras can traverse the cameras by remote control to focus on the nozzle surfaces. The frame rate of the IR camera is 1 fps, the accuracy of its measurement is within $\pm 10^{\circ}C$ and the number of the pixels is 320 $\times$ 240. The accuracy of the K-thermocouples is within $\pm 2.5^{\circ}C$.

Each cooling plenum and the insert of the CMC nozzle were instrumented with thermocouples and pressure probes. The coolant flow rate was adjusted by regulating valves and measured with a calibrated mass flow meter. The temperature data measured by the thermocouples were monitored and stored in PCs every 0.5 sec.

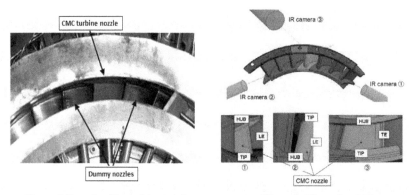

| Figure 7 Cascade of the turbine nozzles | Figure 8 View of each IR camera |

Flow in the cascade

Before cooling performance tests, the velocity distribution in the cascade was calculated by CFD as shown in Figure 9. This study employed Fluent 6.3.26 (Fluent Inc.) as a CFD tool with a hex mesh, and as a turbulence model with the k-ε realizable model selected. Nevertheless, as a wall function, the Non-Equilibrium Wall function was selected. The total cell number was 15,000 points.

Figure 9 indicates the periodicity of the flow field in the cascade. From the computed velocity distribution over the nozzle surface, static pressure fields both at the inlet and the outlet of the cascade, flow rate, and velocity vectors, the flow condition around the CMC nozzle was considered to be equivalent to that in actual engines. Also, in ahead of the cooling test, a preliminary-run of the test rig was conducted to verify the rig functions with a metal nozzle instead of the CMC nozzle.

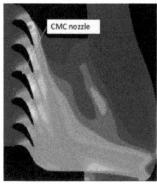

Figure.9 Velocity distribution in the test section (CFD)

Impingement cooling performance test condition

The flow parameters of the impingement cooling test are listed in Table 3. The arrangement of the cooling holes on the insert was designed as similar to that of typical cooled nozzle of engines. One of the important factors of the impingement cooling is coolant flow Reynolds number, Rec. It is defined by the average impingement jet velocity and the impingement-hole diameter. The design point of Re,c of the present CMC nozzle was 1×10^4, which is consistent with those of typical gas turbine engines. In this test, the range of Re,c was varied from 5.73×10^3 to 1.58×10^4 by controlling the coolant flow rate. The test under no coolant condition (Case 6 in Table 3) was also conducted. With the important non-dimensional parameters matched to the engine condition, the actual gas temperature (Tg) and coolant temperature (Tc) were set lower than the engine conditions to minimize measurement error.

Table3 Flow parameters in the impingement cooling performance test

	Case1	Case2	Case3	Case4	Case5	Case6
Re,g	3.8×10^5	3.8×10^5	3.8×10^5	3.8×10^5	3.8×10^5	3.8×10^5
Re,c	1.58×10^4	1.34×10^4	1.19×10^4	8.70×10^3	5.73×10^3	–

With a film-cooled vane, cooling air supplied to the impingement holes are finally bled through the film cooling holes and/or the TE slots. In this test, the CMC nozzle without film holes or TE slots was used to evaluate only the internal cooling performance. To discharge the coolant out of the vane, the insert was notched (Figure 10). As shown in Figure 11, coolant first entered into the insert through the cooling tube (i), then passed through the impingement holes (ii), flowed through the cavity between the insert and the vane shell (iii), and was bled into the casing of the test rig through the notches (iv). Finally, the coolant was discharged into and mixed with the main gas flow. The size of the notches was determined for the impingement cooling to operate at the designed condition.

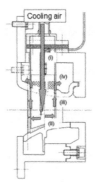

Figure 10 Notch geometry of the insert   Figure 11 Flow path of the cooling air in the insert

Film cooling performance test condition

The flow parameters of the film cooling test are listed in Table 4. As shown in Fig. 12, diameters of the film cooling holes were different by each row and their blowing ratios (BR) were also different. In the table, the minimum and the maximum BR in each case are shown. The operating condition and the configuration of the CMC nozzle were almost identical to that of the impingement cooling test, except that the insert had no notch geometry. In this test, the pressure of the coolant was varied to cover he impingement cooling Reynolds number range in the impingement cooling test.

Before and after the test, pressure-flow characteristic of the CMC nozzle was measured to confirm no leakage. And, it was confirmed that the flow characteristic showed no change between before and after the test.

Table4   Flow parameters in the film cooling performance test

	Case1	Case2	Case3	Case4	Case5	Case6	Case7
Re,g	3.5×10^5	3.5×10^5	3.5×10^5	3.5×10^5	3.5×10^5	3.5×10^5	3.5×10^5
Re,c	1.34×10^4	1.26×10^4	1.16×10^4	1.02×10^4	9.5×10^3	8.4×10^3	7.5×10^3
BR(No.4)	4.6	4.3	3.9	3.4	3.1	2.7	2.3
BR(No.7)	1.2	1.2	1.1	1.1	1.1	1.0	1.0

Figure 12 Film cooling holes on the CMC nozzle

RESULTS AND DISCUSSION

Impingement cooling performance test

The cooling effectiveness as evaluated from the experimental data is shown in Figure 13 along with a prediction. The temperature fields captured by IR cameras were calibrated by the K-thermocouples. The prediction curves were based on a two-dimensional heat-transfer analysis with consideration of some CMC's characteristic such as its anisotropy in thermal conductivity. It was calculated using 1-D finite element method with thermal network. And there are grids along to nozzle surface. And in through-thickness direction, there is one grid. As a boundary condition of main gas side, heat-transfer efficiency based on boundary-layer consumption was used. As a boundary of coolant side, general equation of impingement was used. But correction factors for correlations were not used.

The locations of impingement holes are indicated by arrows in Figure 13 (A).

Some grey shaded regions in the Figure 13 were not visible with the IR cameras because the regions were hidden by the dummy nozzles or were out of the sights of the cameras. A surface temperature at -81%CaX was measured by a K-thermocouple. The temperature at -81%CaX point (measured by the thermocouple) is most reliable because there was no need of calibration such as with IR cameras. In three cases, at that point, the predicted and the measured cooling effectiveness agree well. And as the coolant rate varies, both the prediction and the measured curves shift by roughly same degree.

Minor discrepancies between the prediction and the data were due to low accuracy in distance and curvature correction on the nozzle surface and in environmental correction such as adjusting thermal radiation from the other dummy nozzles or the passage duct.

At -50%CaX to -10%CaX, prediction curve is higher than the measurement curve. That is to say, the heat transfer coefficient of impingement jet in the experiment was lower than the prediction. This is explained that the present impingement holes pitch was out of applicable range of the heat-transfer correlation in the prediction.

Film cooling performance test

Experimental data of the cooling effectiveness are shown in Figure 14 along with the prediction. The prediction considered both impingement and film cooling. In Figure 14 (A), the film cooling holes can be seen clearly in the IR cameras pictures.

In case 4 shown in Figure 14(B) and case 7 in Figure 14(C), the predicted cooling effectiveness at -80%CaX agree well with the measured data by the thermocouple. In case 1, however, the measured data deviates considerably from the predicted curve. This is because the BR of the film cooling in this case was much higher (BR=4.5) than the appropriate range (typically BR from 0.5 to 2.5). Thus, the prediction should not be reasonably accurate.

At -50%CaX to -10%CaX, in comparison with the impingement cooling test, disagreement between the prediction and the data is a bit larger. This suggests that the prediction accuracy near the film cooling holes may be lower.

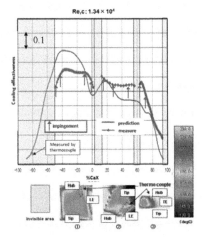

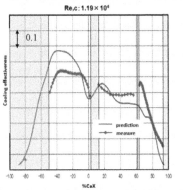

Figure 13(A) Cooling effectiveness (Case 2)

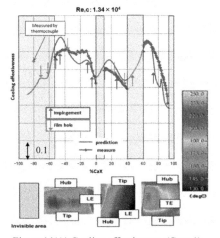

Figure 14(A) Cooling effectiveness (Case 1)

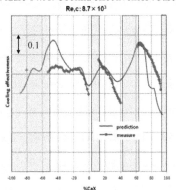

Figure 13(B) Cooling effectiveness (Case 3)

Figure 14(B) Cooling effectiveness (Case 3)

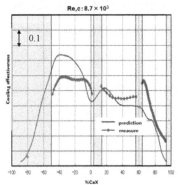

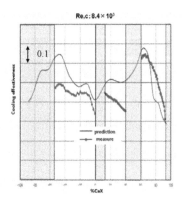

Figure 13(C) Cooling effectiveness (Case 4) | Figure 14(C) Cooling effectiveness (Case 6)

CONCLUSIONS

This paper presents the cooling performance of a CMC nozzle vane in a cascade rig test. The results are compared with prediction. Two types of cooling performance tests were conducted, that is, the one with only the impingement cooling and the other with both impingement and film cooling. Important findings obtained in this study are summarized as follows.

1. With the film cooling performance, the predicted cooling effectiveness is higher than the data with lower BR range. This is because the correlation used in the prediction models film cooling such that the air tends to adhere to the surface and the cooling effectiveness is high when BR is low. But, the experimental result shows some contrasting behavior. This phenomenon needs to be further investigated in the future work.

2. With the impingement cooling performance, the prediction shows fair agreement with the experimental data in the suction surface for all the test cases. However, the prediction is higher than the data in the pressure side region. The reason considered is that the hole arrangement was out of the valid range of the correlation used in the prediction. CMC can withstand in higher temperature than a metal, thus weaker cooling is needed for a cooling CMC airfoil. Thus longer X/d arrangement must be needed in the CMC cooling techniques. This study indicated that new correlation equation to predict the heat transfer coefficient for the long X/d impingement cooling is needed for design CMC cooling airfoil.

ACKNOWLEDGMENTS

The authors would like to express their thanks to the Ministry of Economy, Trade and Industry (METI).

NOMENCLATURE

%CaX   Axial location in percentage to the whole nozzle axial chord length
%Span   Height location in percentage to the whole nozzle passage height
%Wc   cooling flow rate to mainflow rate ratio
BR   Blowing ratio ($BR = \dfrac{\rho_c v_c}{\rho_g v_g}$)
Ca   Chord length of nozzle
LE   Leading edge of airfoil
P   Static pressure
PS   Pressure side

PTF   Pattern factor, Eq.(1)
Re   Reynolds number ($Re = \dfrac{\rho V C a}{\mu}$)
SS   Suction side
Sec   Second
t   Time
TE   Trailing edge of airfoil
T   Temperature
V   Velocity
W   Flow rate

Greek symbols

μ   Gas viscosity
η Cooling effectiveness, Eq.(2)
ρ Density

Subscripts

ave   Average
c   Coolant
g   Main flow
in   At the inlet
w   On the nozzle surface

REFERENCES

[1] F. Zhong and G.L. Brown, "Experimental study of multi-hoe cooling for integrally-woven," International Journal of Heat and Mass Transfer 52 (2009) pp. 971-985.

[2] D.N. Brewer et al., "Ceramic matrix composite vane subelement burst testing," GT2006-90833, ASME Turbo Expo 2006, May 2006.

[3] V. Vedula et al., "Ceramic matrix composite turbine vanes for gas turbine engines," GT2005-68229, ASME Turbo Expo 2005, June 2005.

[4] M. Verrilli and A.Calomino,"Ceramic matrix composite vane subelement testing in a gas turbine environment,"GT2004-53970, ASME Turbo Expo 2004, June 2004.

[5] T. Nakamura, "Development of a CMC   Thrust Chamber," 23rd Annual Cocoa Beach, Conference and Exposition on Advanced Ceramics and   Composites   Volume 20 Issue 4 (1999) pp. 39 −46

[6] H. Murata, et al., IHI Engineering Review, Vol.46, Number3, 2006, 101-108 (Japanese)

[7] K. Nita et al., "Annular cascade rig tests of a CMC stator vane with cooling structure,"IGTC2011-ABS-0062, IGTC'11, November 2011

STUDY ON STRENGTH PREDICTION MODEL FOR UNIDIRECTIONAL COMPOSITES

Hongjian Zhang, Weidong Wen, Haitao Cui, Hui Yuan, Jianfeng Xiao
 College of Energy & Power Engineering, Nanjing University of Aeronautics and Astronautics, 210016 Nanjing, Jiangsu, P.R. China
State Key Laboratory of Mechanics and Control of Mechanical Structures, Nanjing University of Aeronautics and Astronautics, 210016 Nanjing, Jiangsu, P.R. China

ABSTRACT

The fibers in unidirectional composites were treated at two levels: fiber bundle and array of fiber bundles. The fibers were assumed to be placed in hexagonal arrays in the fiber bundle model. And the fiber bundles were assumed to be placed in planar array in unidirectional composites. Based on the statistical theory of crack evolution and the perfect evolution process of random crack cores, the evolution probability of a random crack core with consideration of the influence of length was given, and a two-level model of random crack cores for forecasting the longitudinal tensile strength of unidirectional composites was built. In this two-level model, the average stress concentration factor of the bundle and the stress concentration factor of the fiber inside bundle were calculated respectively. Behavior of various composites was predicted by the model. Comparisons show that several predicted values fit well with the experimental results, but there are gaps between the experimental and predicted values for other materials. The reason for this difference is discussed in this paper.

INTRODUCTION

Composite materials are finding increasing application in industry due to their high strength-to-weight ratio. Typical applications include: aerospace, marine, motor sport, military and specialty construction. The strength of a unidirectional composite plays one of the most important roles in structural design that uses composite laminates because it generally governs the final failure of the structural materials. There are numerous theoretical analyses of the strength of unidirectional composites via micro-stress analysis combined with micro-strength criterions. However, they are always unsatisfying because of capability dispersing of constituent materials [1].

During the past several decades, the Global Load Sharing (GLS) model has been applied widely. The well-known Gucer-Gurland-Rosen chain model is proposed based on the weakest link theory by Gucer and Gurland [2] and Rosen [3]. In this model, a unidirectional composite is treated as a series of chains of lengths equal to the ineffective lengths, and the failure of one chain means the invalidation of the whole composite. Nevertheless, the global load sharing principle is not deemed to be reasonable. The fragmentation model was proposed based on the shear-lag theory by Curtin [4]. In this model, the density of breaks was described by considering load redistribution near the fiber breakpoints. It was indicated that accretion of the break density is the direct reason of a gradual stiffness decrease and final break down of composites. Although the density of breaks was analyzed in the model, congregative effect was ignored because of adopting the global load sharing principle.

As a more practical model, the local load sharing (LLS) model has been widely studied. Zweben [5] proposed a statistical theory of crack evolution by adopting the local load sharing principle for analysing stress concentrations around cracks. The higher local stress increases the break probabilities of intact fibers around cracks. Even if the average stress level of composites is not enhanced, it is possible that intact fibers around cracks break continuously until final break down of whole composites under the effect of local stress concentrations.

This statistical crack evolution theory was developed by Batdorf [6, 7]. It was considered that a composite failure will occur when a crack with some broken fibers is self-propagating. A random critical-core model for predicting the strengths of unidirectional composites was proposed by Zeng [8]. According to the characteristics of crack propagation, the crack ineffective length increases with the number of broken fibers, which is deemed to be one of the main achievements in the model. However, unidirectional composites were assumed impractically to be plane lamellas with two-dimensional fiber arrays.

In this paper, fibers in unidirectional composites were treated at two levels: fiber bundle and array of fiber bundles. The fibers were assumed to be placed in hexagonal arrays in the fiber bundle. And the fiber bundles were assumed to be placed in planar array in unidirectional composites. On the basis of the crack propagation rules, the numerical relationship of the number of random crack cores, the evolution probability of a random crack core evolving to critical size and the failure probability of a composite was deduced, and a theory of random crack cores was presented. And a perfect evolution process of random crack cores was proposed based on the local load sharing (LLS) principle. Then, a two-level random crack core model for predicting the strengths of unidirectional composites was founded. At last, the model was used to predict behaviour of various unidirectional composites.

RANDOM CRACK CORE THEORY

Due to the strength dispersion of single fibers, some weak fibers may break first when a composite is subjected to a finite load, and then some random crack cores germinate in the composite. Intact fibers around random crake cores maybe break due to stress concentrations. When the size of a random crack core is big enough, the random crack core will be self-propagating, and then the whole composite will fail ultimately.

The primary contents of the random crack core theory are as follows: (1) the failure probability of a composite under a certain stress is equal to the probability of an existing crack core evolving to critical size; (2) the number of random crack cores equals to the number of breaches germinated originally, and these crack cores distribute randomly and independently in the composite; (3) if the further evolution probability of a random crack core is large enough, its further evolution could be treat as an inevitable event, and the size of the random crack core at that moment is the critical size; (4) the event of a random crack core evolving to the critical size means the failure of the whole composite.

NUMBER OF RANDOM CRACK CORES

Strengths of single fibers are variable, which was proved following two-parameter Weibull distribution by Pheonix [9]. The expression of strength distribution of single fibers is given as follows:

$$F(\sigma_f, \delta) = 1 - \exp\left[-\frac{\delta}{L_0} \left(\frac{\sigma_f}{\sigma_0} \right)^{\beta} \right] \qquad (1)$$

Where σ_f is the tensile stress of fibers, δ is the length of fiber; σ_0 is scaling parameter corresponding to fiber length L_0; β is shape parameter.

The number of random crack cores in a composite equals to the number of breaches germinated originally, and these crack cores distribute randomly and independently in the composite. If mechanical property of fibers is unambiguous, the density of random crack cores, i.e. the number of random crack cores in a united length fiber, is determined only by the average tensile stress of fibers and moreover increases with the average tensile stress.

When the average tensile stress of fibers is σ_f, the density of random crack cores ρ is given as follows:

$$\rho = \frac{F(\sigma_f, \delta_0)}{\delta_0} \tag{2}$$

Here δ_0 is the ineffective length of composites.

The ineffective length δ_0 of composites is determined jointly by characters of fibers, matrix and interface, which could be calculated by the shear-lag model. With a hypothesis that tensile stress of broken fibers is linear with the distance from breakpoint in the stress-recovery zones, an approximate expression on the ineffective length δ_0 is given based on Kelly-Tyson shear-lag model [9] as follows:

$$\delta_0 = 2 \times \frac{\pi(d/2)^2 \sigma_f}{\pi d \tau_s} = \frac{d\sigma_f}{2\tau_s} \tag{3}$$

Where d is the diameter of fibers; τ_s is the strength of fiber-matrix interface.

The number of random crack cores m equals to the product of the density of random crack cores and the total length of fibers V, which could be expressed as

$$m = Int[\rho V] = Int[\frac{LN}{\delta_0} F(\sigma_f, \delta_0)] \tag{4}$$

Here, $Int[\ ]$ is the function of getting integers; L is the length of a composite; N is the total of fibers in the composite; the total length of fibers in the composite $V = LN$.

CRITICAL SIZE OF RANDOM CRACK CORES

As the fibers in unidirectional composites were treated through fiber bundle level and array of fiber bundles level, there are three different cases of critical crack cores: (1) several bundles and several fibers; (2) several bundles; (3) several fibers in one bundle. And the critical size can be calculated by :

$$F\left(K_{N_{cri}-1}\sigma_f, \delta_{N_{cri}-1}\right) < 1-e^{-1} \leq F\left(K_{N_{cri}}\sigma_f, \delta_{N_{cri}}\right) \tag{5}$$

Where, $K_i = \max\left(K_{i,j}\right), i < j \leq n_{con}$, and $K_{i,j}$ is the stress concentration factor of the fiber whose serial number is j when i fibers have broken in the random crack core; n_{con} is the biggest serial number of fibers enduring concentrative stresses when i fibers have broken in the random crack core; δ_i is the influenced-length of the random crack core with i broken fibers.

CALCULATION OF STRESS CONCENTRATION FACTOR

According the study of Pheonix [11], there are two types of stress concentrations: (1) the average stress concentration among fiber bundles; (2) the stress concentration among fibers in one bundle.

Based on the assumption that every bundle can be regarded as one union, the average stress concentration among fiber bundles can be calculated by:

$$K_i = \sqrt{\frac{\pi i}{4} + 1} \tag{6}$$

As the fibers were assumed to be placed in hexagonal arrays in fiber bundle, the stress concentration among fibers in one bundle can be calculated by:

$$K_j = \sqrt{\frac{2\sqrt{j}}{\pi^{3/2}} + 1} \qquad (7)$$

ANALYSIS ON EVOLUTION PROBABILITIES OF RANDOM CRACK CORES

Evolution of a random crack core is an indeterminate process, and evolution paths are various. In this paper, a perfect evolution process of a random crack core was presented on the assumptions that the fiber whose stress concentration factor is the largest will be the next broken fiber, and the failure probability of the fiber equals to the probability of any fiber around the random crack core breaking in a certain state. If stress concentration factors of some fibers are the coordinately largest, the nearest fiber from the center of the random crack core (the original crack core) is enacted to break next. If some of these fibers from the center of the random crack core are the coordinately nearest, the first fiber in counter-clockwise direction from the last broken fiber in these fibers is enacted to break next. According to aforementioned rules and combining a load sharing principle, every evolution step of a random crack core is unambiguous.

Combining the local load sharing principle, evolution steps of a random crack core are determined. According to the sequence of failure, fibers are marked in turn, just as fig1.

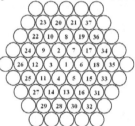

Fig1 The perfect evolution process of a random crack core

As a particular path of the Markov process, the perfect evolution process of a random crack core is not only a path with the largest probability, but also a unique path insuring every evolution step of the random crack core being a steady state. It is determined that if an evolution step of a random crack core departs from corresponding steady state, the following steps will approach the steady states with much larger probabilities.

When the number of broken fibers in a random crack core is i, the failure probability of the no. j fiber ($j > i$) under concentrated stress is shown as follows

$$p_{i,j}(\sigma_f) = \frac{F(K_{i,j}\sigma_f, \delta_i) - F(K_{i-1,j}\sigma_f, \delta_{i-1})}{1 - F(K_{i-1,j}\sigma_f, \delta_{i-1})} \qquad (8)$$

Here, $K_{i,j}$ is the stress concentration factor of the no. j fiber when the number of broken fibers in a random crack core is i, which could be calculated according to the local load sharing principle.

According to the assumptions of the perfect evolution process of random crack cores, when the number of broken fibers in a random crack core is i, the probabilities of the number of broken fibers from i to $i+1, i+2$, and $i+3$ in a step, marked respectively as $p_{i \to i+1}$, $p_{i \to i+2}$ and $p_{i \to i+3}$, are calculated as follows

$$p_{i \to i+1}(\sigma_f) = \sum_{j=i+1}^{n_{con}} \frac{p_{i,j}}{1 - p_{i,j}} \prod_{j=i+1}^{n_{con}} (1 - p_{i,j}) \qquad (9.1)$$

$$p_{i \to i+2}(\sigma_f) = \sum_{j=i+1}^{n_{con}-1} \sum_{k=j+1}^{n_{con}} \frac{p_{i,j} p_{i,k}}{(1-p_{i,j})(1-p_{i,k})} \prod_{j=i+1}^{n_{con}} (1-p_{i,j}) \tag{9.2}$$

where n_{con} is the biggest serial number of fibers enduring concentrative stresses at the moment. When the number of broken fibers in a random crack core is i, the probability of the number of broken fibers from i to $i+q$ in a step, which is marked as $p_{i \to i+q}$, can be deduced by analogy.

The evolution probability of a random crack core with i broken fiber is shown as follows

$$p_i(\sigma_f) = \sum_{q=1}^{n_{con}} p_{i \to i+q} = 1 - \prod_{j=i+1}^{n_{con}} (1-p_{i,j}) \tag{10}$$

Chances of some fibers breaking simultaneously exist in the evolution processes of random crack cores, but it is unpractical to consider such circumstances fully because of their complexity. The possibility of some fibers breaking simultaneously was always ignored. Here, multiform algorithms, such as ignoring some fibers breaking simultaneously, considering two fibers breaking simultaneously and considering ω fibers breaking simultaneously, are deduced.

(1) Ignoring some fibers breaking simultaneously

When the stress level is low, the probability of some fibers breaking simultaneously is much smaller than that of fibers breaking one by one. For the sake of ease of the analysis process, the possibility of some fibers breaking simultaneously is always ignored. Then, the evolution possibility of a random crack core from one broken fiber to the critical size r is shown as follows

$$P_r(\sigma_f) = \prod_{i=1}^{r-1} p_i \tag{11}$$

(2) Considering two fibers breaking simultaneously

The algorithm ignoring some fibers breaking simultaneously cannot but bring errors to calculated results of evolution possibilities of random crack cores. The algorithm considering some fibers breaking simultaneously is able to decrease errors. With considering two fibers breaking simultaneously, the evolution possibility of a random crack core from one broken fiber to the critical size r could be calculated as follows

$$P_1(\sigma_f) = 1 \tag{12.1}$$

$$P_2(\sigma_f) = p_{1 \to 2} \tag{12.2}$$

$$P_i(\sigma_f) = p_{i-1 \to i} P_{i-1} + (p_{i-2} - p_{i-2 \to i-1}) P_{i-2} \quad ; \quad 2 < i < r \tag{12.3}$$

$$P_r(\sigma_f) = p_{r-1} P_{n-1} + (p_{r-2} - p_{r-2 \to r-1}) P_{r-2} \tag{12.4}$$

ANALYSIS ON EVOLUTION PROBABILITIES OF TWO-LEVEL RANDOM CRACK CORES

There are two steps for calculating the evolution probabilities of two-level random crack cores[12].

(1) To determine whether the critical random crack cores are started in the first bundle

After the (j-1)th fiber in first bundle is broken, if $K_{j-1}\sigma < \sigma_{\delta_{1j-1}}$, this means that the critical random core has not formed. If j is equal to m (the numbers of fibers in a bundle), the first bundle will be broken. And the broken probabilities can be decided by:

$$W_1(\sigma) = F(\sigma)\left\{1-\left[1-F(K_1\sigma)\right]^{N_1}\right\}\cdots\times\left\{1-\left[1-F(K_{m-1}\sigma)\right]^{N_{m-1}}\right\} \qquad (13)$$

Where $N_j = \sqrt{4\pi j^2}^{\frac{1}{2}}$.

After the (j-1)th fiber is broken, if $K_{j-1}\sigma > \sigma_{\delta_{1,j-1}}$, this means that the size of critical random core is j (Case 3).

(2) To determine whether the critical random crack cores are founded in the ith bundle

When the i-1th bundle is broken and the effective value length is δ_{i0}, the stress of the first fiber in ith bundle is $KK_{i-1}\sigma$. If $KK_{i-1}\sigma > \sigma_{\delta_{i0}}$, the size of critical random core is i-1 bundle (Case 2), and the probability is:

$$P(\sigma) = W_1(\sigma)\prod_{q=2}^{(i-1)-1}\left\{W_q(\sigma)\left[1-W_q(\sigma)\right]\right\}\bullet\left\{1-\left[1-W_{i-1}(\sigma)\right]^2\right\} \qquad (14)$$

When the j-1th fiber in ith bundle is broken and the effective value length is δ_{i0}, the stress of the jth fiber in ith bundle is $KK_{i-1}K_{j-1}\sigma$. If $KK_{i-1}K_{j-1}\sigma > \sigma_{\delta_{i0}}$, the size of critical random core is i-1 bundle + j-1 fibers (Case 1), and the probability is:

$$\tilde{p} = 1-(1-p)^t \qquad (15)$$

Figure 2 shows the flow chart of the two-level random crack core model.

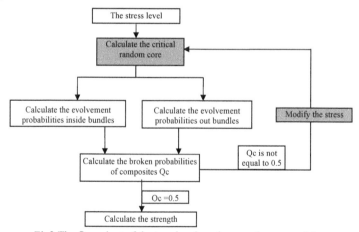

Fig2 The flow chart of the two-level random crack core model

VALIDATION AND DISCUSSION

In order to verify the effectiveness of the two-level model, the strengths of various unidirectional composites are predicted. The experimental results are summarized in Table 1[10], where σ_{UD} is the unidirectional composite strength, σ_0 is fiber strength, m is fiber distribution, τ_i is the interfacial shear strength, σ_m is matrix strength.

Table 1 Materials and experimental results [10]

No.	Material	σ_{UD} MPa	v_f	σ_0 MPa	m	τ_i MPa	σ_m MPa
1	Carbon (M40)/Epoxy	2310	0.6	4500	16	50	80
2	C/C (2600_C heat treated)	852	0.6	3150	4.9	17	5
3	C/C (2900_C heat treated)	1020	0.6	3400	5.8	15	5
4	C/C (Resin char method)	1360	0.6	3700	5.7	17	5
5	C/C (Hot isostatic pressing method)	1080	0.6	3700	5.7	18	5
6	CMC (Curtin's value in Curtin [1])	700	0.5	1700	4	5	20
7	CMC (thick coated interface/ modified PIP matrix)	320	0.2	3600	5	20	6

Table 2 Comparison between prediction and experiments

No.	Material	Experiments σ_{UD} MPa	Prediction σ_{UD} MPa	Error %
1	Carbon (M40)/Epoxy	2310	1980	-14.3
2	C/C (2600_C heat treated)	852	1190	39.7
3	C/C (2900_C heat treated)	1020	1282	25.7
4	C/C (Resin char method)	1360	1399	2.9
5	C/C (Hot isostatic pressing method)	1080	1409	30.5
6	CMC (Curtin's value in Curtin [4])	700	639	-8.7
7	CMC (thick coated interface/ modified PIP matrix)	320	439	37.2

Table 2 lists the comparison between prediction and experiments of various composites. Comparisons show that several predicted values fit well with the experimental results, but there are gaps between the experimental and predicted values for some materials.

The composite tensile strength is decided by the fiber strength distribution, interfacial strength and the matrix strength. Actually, a large number of fibers will fail simultaneously. The number of fibers fail simultaneously will increase with the interfacial strength. And the composite strength will decrease with the interfacial strength. These phenomena can be verified by the experiments listed in Table 1. If the interfacial strength is assumed to be zero, the composite strength must correspond to the dry bundle strength. But the possibility of some fibers breaking simultaneously was ignored in our model due to of the complexity of fibers fail simultaneously. That means, the actual strong interfacial strength is regarded as weak interfacial strength in predicted model. It will cause the predicted value to be greater than the experimental value. That is why there are gaps between the experimental and predicted values for the some materials.

CONCLUSIONS

A two-level crack core model for predicting the strengths of unidirectional composites is developed in this paper. The model, based on the random crack core theory and the perfect evolution process, has been used to predict behaviour of various unidirectional composites. Comparisons show that several predicted values fit well with the experimental results, but there are gaps between the experimental and predicted values for the other materials. The difference is attributed to the possibility of some fibers breaking simultaneously, which was ignored in the model.

REFERENCES

[1] K. Reifsnider, S. Case, J. Duthoit. The mechanics of composite strength evolution, Composites Science and Technology, 2000, 60:2539~2546.
[2] Gucer D. E., Gurland J., Comparison of the statistics of two fracture modes, Journal of the Mechanics and Physics of Solids,1962,10(4):365~373.
[3] Rosen B. W., Tensile failure of fibrous composites, AIAA, 1964, 2:1985~1991.
[4] Curtin W.A., Theory of Mechanical Properties of Ceramic-Matrix Composites, Journal of the American Ceramics Society, 74 (11): 2837~2845.
[5] Zweben C., Tensile failure of fiber composites, AIAA, 1968, 6(12):2325~2331.
[6] Batdorf S. B., Tensile strength of unidirectionally reinforced composites, J. Reinf. Plast. Composites, 1982, 14:153-164.
[7] Batdorf S. B., Ghaffarian R. Tensile strength of unidirectionally reinforced composites—II, J. Reinf. Plast. Composites, 1982, 14:165-176.
[8] Qing-Dun Zeng, A random critical-core theory of microdamage in interply hybrid composites: II ultimate failure, Composites Science and Technology, 1994,52(4):481~487.
[9] Smith R. L., Pheonix S. L. A comparison of probabilistic techniques for the strength of fibrous materials under local load-sharing among fibers, International Journal of Solids and Structures,1983, 19:479-496.
[10] Jun Koyanagi, Hiroshi Hatta, Masaki Kotani, etc. A Comprehensive Model for Determining Tensile Strengths of Various Unidirectional Composites, Journal of Composite Materials, 2009, 43(18):1901-1914.
[11] Sivasambu Mahesh, S. Leigh Phoenix, Irene J. Beyerlein. Strength distributions and size effects for 2D and 3D composites with Weibull fibers in an elastic matrix, International Journal of Fracture, 2002, 115: 41~85.

[12] Zhang Lijun, Wang Deyu, Zhu Xi, Zhang Lingjian, etc. Strength prediction for unidirectional FRP based on the theory of two-leveled randomly enlarging critical core. Acta Materiae Composite Sinica. 2010, 27(2): 140-147.

Processing and Properties of Fibers and Ceramics

THE EFFECT OF THE ADDITION OF CERIA STABILISED ZIRCONIA ON THE CREEP OF MULLITE.

D. Glymond*, M. Vick°,+, M.-J. Pan°, F. Giuliani*,+, L.J. Vandeperre*

*Centre of Advanced Structural Ceramics & Department of Materials, Imperial College London, South Kensington Campus, London SW7 2AZ, UK

+Department of Mechanical Engineering, Imperial College London, South Kensington Campus, London SW7 2AZ, UK

° Naval Research Laboratory, Washington, DC 20375

ABSTRACT

Mullite is considered a promising candidate for ceramic recuperators in turbo propelled engines, due to due to its low thermal conductivity, adequate thermal shock resistance, low cost, low density, thermodynamic stability, and reasonable strength at high temperatures. Unfortunately, the limited fracture toughness of mullite ($\sim$1.8-2.8 MPa m$^{1/2}$) is considered too low. Improving the fracture toughness to 4.7 MPa m$^{1/2}$ is possible by the addition of ceria stabilised zirconia (CSZ). However, the addition of CSZ to mullite may also affect other properties. In this paper the effect of the addition of CSZ on the creep resistance is described by comparing a mullite-zirconia composite made with a commercially available mullite powder against the creep behaviour of mullite made from the same mullite powder. The stress exponent is close to 1 and the activation energies for creep were similar at 426±38 kJ mol^{-1} and 452±15 kJ mol^{-1}. However, the resistance to creep as expressed by a given strain rate for a given stress is 250 °C lower in the 20 vol% zirconia-mullite composite compared to the baseline.

INTRODUCTION

Mullite is an important material for high temperature structural applications due to its low coefficient of thermal expansion, good creep resistance and strength at these temperatures.[1-2] The main drawback of mullite as a structural material is its fracture toughness of only 1.8-2.8 MPa m$^{1/2}$.[3-4]

There are a variety of methods to increase the toughness of a ceramic such as crack deflection[5], self-reinforcement by elongated grains to create crack bridging,[6] and second phase particles.[7] One method, which has been shown to toughen mullite to a large extent, is the addition of zirconia. Bender et al. measured a toughness of 4.7 MPa m$^{1/2}$ when adding 18 vol% of ceria-stabilised zirconia (CSZ).[8] With toughness values in this range a number of structural applications within turbine engines are within reach, and thus the effect of this addition on other properties needs to be quantified to determine overall feasibility.

The high creep resistance of mullite is well known[9-13], and any effect on this property will have a large effect on the performance of the material.[1, 2, 14] Therefore in this paper the effect of the addition of CSZ on the resistance to compressive creep is investigated.

EXPERIMENTAL

A commercial mullite powder (KM 101, KCM corporation, Nagoya, Japan) with a nominal composition of 28 wt% SiO$_2$ and 71.9 wt% Al$_2$O$_3$ and less than 0.1 wt% other oxides was used. This powder is very close to the silica rich end of the single phase mullite region represented by the nominal composition 3 Al$_2$O$_3$ – 2 SiO$_2$. The Ce-stabilised zirconia contained 12 mol% CeO$_2$ (Grade CSZ-12) and was obtained from Iwatani America, USA.

Composites containing 20 vol% CSZ were prepared by ball milling the powder mixture in acetone for 5 hours. Poly-ethylene glycol (PEG 400, Alfa Asaer, UK) was added as a binder. After milling, the slurries were dried in an oven at 70 °C overnight followed by uni-axial compaction at

30 MPa before sintering in air at 1500 °C for 5 hours. Samples of the mullite powder on its own were produced by vacuum hot pressing the powder at 1650 °C for 2 hours under 25 MPa pressure.

The density of the sintered pellets was determined using Archimedes' principle. The theoretical densities were calculated using the rule of mixtures using 3.16 Mg m^{-3} for mullite and 6.23 Mg m^{-3} for Ce-stabilised zirconia[15]. The microstructure was characterised by observing fracture surfaces in a scanning electron microscope (JEOL-5610LV, Jeol, Japan).

All samples for creep tests were produced in a 2:1 aspect ratio. For the reference material consisting of mullite alone, samples with dimensions of 5 mm x 5 mm x 10 mm were machined out of the hot pressed billet, while for the mullite-zirconia composites, cylindrical samples measuring 12 mm high by 6 mm diameter were produced by sintering directly. All samples were polished plane parallel for compressive creep testing in a custom built testing rig consisting of a MRF Vacuum furnace mounted in a 100 kN universal test frame with graphite push rods and molybdenum heat shields and elements. The samples were heated under vacuum to the test temperature at 50 °C min^{-1}, soaked at temperature for 20 minutes under a pinch load of 150 N and then tested under load for 1 hour unless failure occurred earlier. After testing the CSZ-mullite composites at temperatures (~1100-1300 °C) and stresses (10-100 MPa) relevant to some of the gas turbine applications, initial tests indicated that the creep rate in the hot pressed mullite was much lower. Therefore, it was decided to find the temperature difference needed to obtain the same creep strain rate for the same stress for both types of mullite, and therefore subsequent testing on the hot pressed mullite was carried out at higher temperatures (1400-1550 °C).

RESULTS

Table 1 shows the density of the samples tested. Both sets of samples were near full density after processing. Figure 1 shows micrographs of the fracture surface of the two materials. The mullite + CSZ sample shows bright equiaxed zirconia grains embedded in a matrix of gray, elongated, mullite grains and evidence of intergranular fracture, whereas the hot pressed mullite shows mainly transgranular failure.

Table 1. Density of the different materials

Material	Density (%)
Hot pressed mullite	98.8 ± 0.2
Mullite + 20 vol% CSZ	99.9 ± 0.2

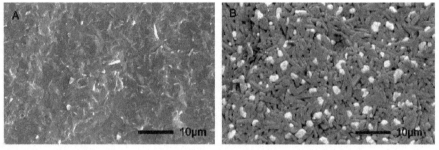

Figure 1. (a) Secondary electron micrograph of the hot pressed mullite, (b) back scattered electron micrograph of the mullite + 20 vol% CSZ composites. Grey areas are mullite, brighter areas are CSZ.

Figure 2 shows examples of typical traces of strain versus time. The initial creep rate tends to be high (primary creep), followed by a stable, slower, steady state creep rate (secondary creep) until the material starts to fail and the creep rate enhances again as can be seen for example in the trace for 52.8 MPa in Figure 2. Hence, to calculate the strain rate, the initial higher strain rate is ignored and the slope of the linear section is determined. These steady state strain rates have been plotted against stress for the mullite-zirconia composite in Figure 3. Over the limited range of stresses tested, the creep rate is related to the stress in the usual manner through:

$$\varepsilon^{\cdot} = A \cdot \sigma^{n} \tag{1}$$

where $\varepsilon^{\cdot}$ is the strain rate, A is a pre-exponential constant, σ is the stress and n the stress exponent. The fact that most lines are parallel, indicates that the stress exponent varies little. As shown in Table 2, this is indeed the case with the stress exponent close to 1 for most temperatures. It is worth pointing out that it is clear from Figure 3 that even at 1300 and 1350 °C the stress exponent would have been closer to 1 had the data point at 52.8 MPa not been included in the analysis.

Figure 4 shows the same strain rate against stress plot for the hot pressed mullite. Again parallel lines were found and as illustrated in Table 2, the stress exponent is again nearly constant and close to 1. Moreover, the results have been arranged in Table 2 to illustrate that for approximately the same creep rate at the same stress, the temperature has to be 250 °C higher for the hot pressed mullite than for the composite containing CSZ.

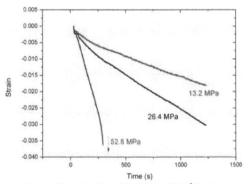

Figure 2. Strain against time for mullite with 20 vol% CSZ at 1300 °C at 3 stress levels.

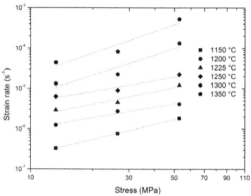

Figure 3. Strain rate versus applied stress the mullite + 20 vol% CSZ composite

Table 2. Strain rate and stress exponent for different temperatures for the hot pressed mullite and the zirconia-mullite composite.

KM + 20 vol% CSZ			Hot pressed mullite		
Temperature (^{o}C)	Strain rate at 13.2 MPa	Stress Exponent	Temperature (^{o}C)	Strain rate at 13.2 MPa	Stress Exponent
1150	$3.3 \ 10^{-7}$	1.3	1400	$5.9 \ 10^{-7}$	1.1
1200	$1.3 \ 10^{-6}$	0.9	1450	$1.6 \ 10^{-6}$	1.1
1225	$2.9 \ 10^{-6}$	1.0	-	-	-
1250	$6.3 \ 10^{-6}$	0.9	1500	$5.4 \ 10^{-6}$	1.2
1300	$1.3 \ 10^{-5}$	1.7	1550	$1.3 \ 10^{-5}$	1.2
1350	$4.4 \ 10^{-5}$	1.8	-	-	-

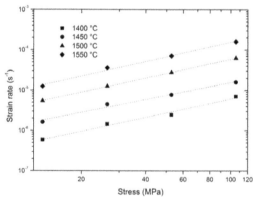

Figure 4. Strain rate versus stress for the hot pressed mullite.

As can be seen in Figure 5, the strain rate data for fixed stress clearly also follows the expected Arrhenius relationship:

$$\varepsilon = B \cdot \exp\left(-\frac{Q}{RT}\right)$$

(2)

where Q is the activation energy, B is a pre-exponential factor, R is the gas constant and T is temperature. The activation energies have been tabulated in Table 3. At lower stress the activation energy is of the order of 400-460 kJ mol^{-1}, but an increase in stress causes the activation energy to increase with values up to 640 kJ mol^{-1} being observed.

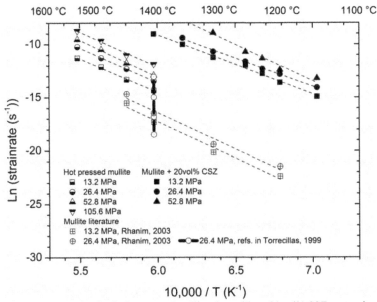

Figure 5. Arrhenius plot for the hot pressed mullite, the mullite + 20 vol% CSZ composite and comparison with literature data from[11, 16].

Table 3. Compressive creep activation energies for different applied stresses and average value (± standard deviation) for stresses below 52 MPa.

Stress (MPa)	Activation Energy Q (kJ mol)	
	KM Mullite	KM Mullite + 20 vol% CSZ
13.2	399	462
26.4	453	441
52.8	533	638
105.6	582	n/a
Average (σ<52 MPa)	426±38	452±15

DISCUSSION

By adapting the temperature, similar creep rates for the same applied stress have been obtained for the compressive creep in the mullite-zirconia composite and the hot pressed mullite and it is clear that the mullite-zirconia composite experiences the same creep rates for the same stress when the temperature is 250 °C lower. This result in terms of reduced creep resistance, agrees rather well with the observation that material transport in general is enhanced in the composite as it can be sintered at 1500 °C, whereas pure mullite is typically sintered around 1700 °C[13].

Indications that the creep mechanism might be similar despite the temperature difference are that the stress exponent is close to 1 for both materials and that on average the activation energy for creep at stresses below 52 MPa is very similar: 426 ± 38 kJ mol^{-1} and 452 ± 15 kJ mol^{-1}.

A stress exponent close to 1 is in line with what has been reported for mullite and the normal interpretation is that it indicates diffusion controlled creep[1, 2]. The tendency for the stress exponent to become larger when higher stresses are applied has also been observed before: Torrecillas et al.[16] argue that the increase arises from slow crack growth and other creep damage evolving rapidly at such stresses. A similar but slightly different explanation is due to Rhanim et al.[11], who suggest that when the stress is high, failure occurs before secondary creep is established and therefore the observed strain rates are too high. That the stress is *"high"* for the experiments where the stress exponent rises can be seen from the very limited creep life (< 10 minutes) at the higher end of applied stresses, see Figure 2.

In the literature, the measured activation energies are either around 400 kJ mol^{-1} or in the range of 700-1000 kJ mol^{-1}, with the latter values normally measured at higher temperatures (>1300 °C) and higher stresses[16]. Since in our experiments, higher activation energies tend to go hand in hand with early creep failure, it is unlikely that the higher values obtained here can be taken as true values. Therefore the activation energies are estimated as 426 ± 38 kJ mol^{-1} and 452 ± 15 kJ mol^{-1}, consistent with the lower value observed in the literature. This lower value is very similar to the activation energy for oxygen diffusion in mullite[12], which has been measured as 397 kJ mol^{-1} [17] and 433 ± 21 kJ mol^{-1} [18] by two independent methods. Hence in line with existing interpretations, the mechanisms could be grain boundary sliding accommodated by diffusion with oxygen diffusion in mullite being the rate determining step of the accommodation process. As illustrated with the data collected by Torrecillas[16] for creep at 1400 °C in Figure 5, our data for the hot pressed mullite falls within the scatter band of measured creep rates for mullite. Therefore a priori there is no reason to believe that the same mechanism would not apply. Torrecillas and co-workers[16] attribute the wide variation in creep rates in the literature to differences in grain size, although other effects such as impurities and porosity could also play a role.

Since the activation energy and stress exponent for creep in the composite with zirconia are the same, it is tempting to conclude that the same rate determining mechanism operates in this material too. However, since the proposed rate determining step is diffusion of oxygen through mullite, it is not obvious how the same process could lead to such disparaging creep rates especially since the grain size does not appear to be especially small. Hence, either the mechanisms for creep in the two materials are different and the agreement in activation energies and stress exponents is coincidental, or the same mechanism operates in both materials but it is not grain boundary sliding limited by oxygen diffusion through the mullite grains.

Indeed, a stress exponent of 1 is also consistent with creep accommodated entirely by viscous flow of the glassy material from the compressive to the tensile surfaces of the grains with creep rates given by[12, 19, 20]

$$\dot{\varepsilon} = \frac{1}{\sqrt{3}} \frac{\sigma}{\eta} \left(\frac{\delta}{d} \right)^3 \tag{3}$$

where σ is the applied stress, η is the viscosity, δ is the thickness of the grain boundary film and d is the grain size. Following de Arellano-Lopez et al.[12], who point out that grain boundary films tend to be of the order of nanometres and grain sizes of the order of microns, the ratio of the grain boundary film thickness to grain size can crudely be estimated to be 10^{-3}. This assumption allows calculating the apparent viscosity of the grain boundary glassy phase from the creep strain rate. The result, see Figure 6, shows that the estimated viscosity of the grain boundary glassy phase for the hot pressed mullite is in between measured viscosities for a glass containing 80 mol% SiO_2 and 20 mol% Al_2O_3 and for a glass containing 94 mol% SiO_2 and 6 mol% Al_2O_3. The latter composition is very close to the composition of the liquid in equilibrium with mullite at the processing temperature of 1650 °C. Moreover, the change in viscosity of the two glasses and of the viscosity estimated from the creep strain rate with reciprocal temperature is very similar. Hence, simple viscous flow is certainly possible as the mechanism of creep in the hot pressed mullite.

The estimated viscosity of the glassy phase in the mullite+zirconia composite is 3 orders of magnitude lower than that of the hot pressed mullite and again shows a reasonable variation against reciprocal temperature. Hence, if it is assumed that the same mechanism applies, the explanation of the difference in creep resistance has to be that either zirconia, ceria or both, dissolve in the grain boundary phase and reduce its viscosity or that the amount of glassy phase is increased dramatically. That such dramatic changes in viscosity are possible for fairly limited changes in composition is in fact illustrated by the difference in viscosity of the two glasses.

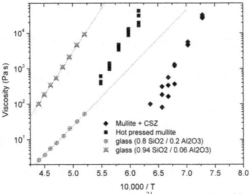

Figure 6. Literature data for the viscosity of Al_2O_3-SiO_2[21] glasses together with estimated viscosity of the glassy phase from the creep strain rate for the hot pressed mullite and for the mullite – Ce-stabilized zirconia composite.

Until transmission electron microscopy has confirmed the presence of sufficient glass on the grain boundaries and the dissolution of ceria or zirconia in the glassy phase, this is proposed as a possible mechanism. Compared to creep being controlled by oxygen diffusion, it has the advantage of offering an explanation for the reason why the creep resistance is lower in the composite, whereas creep controlled by oxygen diffusion does not. It is, however, also possible, that the viscous mechanism is responsible for the high creep rates in the composite whereas the lower creep rates in the hot pressed material are due to grain boundary sliding accommodated by oxygen diffusion through mullite.

CONCLUSIONS

The addition of ceria-stabilized zirconia to mullite, while beneficial for the toughness, reduces the resistance to creep relative to samples made from the same mullite powder without such additions. The same creep rate at the same stress requires 250 °C less in the composite compared to the pure reference material.

In both materials, the stress exponent was found to be close to 1 and the creep activation energies were similar at 426 ± 38 kJ mol^{-1} and 452 ± 15 kJ mol^{-1}. These results are in line with literature values. The increase in stress exponent with temperature and of the activation energy with stress is attributed to steady-state strain rates never establishing because the samples transit from primary to ternary creep without a stable secondary creep regime.

Since it is difficult to explain several orders of magnitude difference in creep strain rate if the rate determining step is the same (diffusion of oxygen through mullite), it is proposed that the creep mechanism is viscous flow of the glassy grain boundary film from the compressive to the tensile boundaries of the grains. It is proposed that the addition of Ce-stabilized zirconia, enhances the creep rates due to either zirconia or ceria dissolving in the glass phase and reducing its viscosity. Further work including transmission electron microscopy is needed to test this hypothesis.

ACKNOWLEDGEMENTS

DG, FG and LV thank the US Office of Naval Research and the Office of Naval Research Global for funding this work through grant N62909-10-1-7083.

REFERENCES

1. P.A. Lessing, R.S. Gordon, and K.S. Mazdiyasni, Creep of Polycrystalline Mullite. *Journal of the American Ceramic Society*, **58**(3-4): p. 149-149 (1975).
2. P.C. Dokko, J.A. Pask, and K.S. Mazdiyasni, High-Temperature Mechanical-Properties of Mullite under Compression. *Journal of the American Ceramic Society*, **60**(3-4): p. 150-155 (1977).
3. S. Kanzaki, H. Tabata, T. Kumazawa, and S. Ohta, Sintering and Mechanical-Properties of Stoichiometric Mullite. *Journal of the American Ceramic Society*, **68**(1): p. C6-C7 (1985).
4. T.I. Mah and K.S. Mazdiyasni, Mechanical-Properties of Mullite. *Journal of the American Ceramic Society*, **66**(10): p. 699-703 (1983).
5. K.T. Faber and A.G. Evans, Crack Deflection Processes - I. Theory. *Acta Materialia*, **31**(4): p. 565-576 (1983).
6. R.W. Steinbrech and O. Schmenkel, Crack-Resistance Curve of Surface Cracks in Alumina. *Journal of the American Ceramic Society*, **71**(5): p. C271-C273 (1988).
7. D.J. Green, Fracture Toughness Predictions for Crack Bowing in Brittle Particulate Composites. *Journal of the American Ceramic Society*, **66**(1): p. C4-C5 (1983).
8. B.A. Bender and M.J. Pan, Selection of a Toughened Mullite for a Miniature Gas Turbine Engine. *Mechanical Properties and Performance of Engineering Ceramics and Composites Iv*, **30**(2): p. 167-175 (2010).
9. L. Falk, J. Pitchford, W. Clegg, E. Liden, E. Carlstrom, and S. Gustafsson, Development of microstructure during creep of polycrystalline mullite and a nanocomposite mullite/5 vol.% SiC. *Journal of the European Ceramic Society*, **29**(4): p. 539-550 (2009).
10. W.M. Kriven, J.W. Palko, S. Sinogeikin, J.D. Bass, A. Sayir, G. Brunauer, H. Boysen, F. Frey, and J. Schneider, High temperature single crystal properties of mullite. *Journal of The European Ceramic Society*, **19**(2529-2541)(1999).
11. H. Rhanim, C. Olagnon, G. Fantozzi, and A. Azim, High-temperature deformation of mullite and analysis of creep curves. *Journal of Materials Research*, **18**(8): p. 1771-1776 (2003).
12. A.R. de Arellano-Lopez, J.J. Melendez-Martinez, T.A. Cruse, R.E. Koritala, J.L. Routbort, and K.C. Goretta, Compressive creep of mullite containing Y_2O_3. *Acta Materialia*, **50**: p. 4325-4338 (2002).

13. R. Torrecillas, G. Fantozzi, S. de Aza, and J.S. Moya, Thermomechanical behaviour of mullite. *Acta Materialia*, **45**(3): p. 897-906 (1997).

14. H. Ohira, H. Shiga, M.G.M.U. Ismail, Z. Nakai, and T. Akiba, Compressive Creep of Mullite Ceramics. *Journal of Material Science Letters*, **10**: p. 847-849 (1991).

15. M. Yashima, T. Hirose, S. Katano, and Y. Suzuki, Structural changes of ZrO_2-CeO_2 solid solutions around the monoclinic-tetragonal phase boundary. *Physical Review B*, **51**(13): p. 8018-8025 (1995).

16. R. Torrecillas, J.M. Calderon, J.S. Moya, M.J. Reece, C.K.L. Davies, C. Olagnon, and G. Fantozzi, Suitability of mullite for high temperature applications. *Journal of The European Ceramic Society*, **19**: p. 2519-2527 (1999).

17. Y. Ikuma, E. Shimada, S. Sakano, M. Oishi, M. Yokoyama, and Z. Nakagawa, Oxygen self-diffusion in cylindrical single crystal mullite. *Journal of the Electrochemical Society*, **146**(12): p. 4672-4675 (1999).

18. P. Fielitz, G. Borchardt, H. Schneider, M. Schmucker, M. Wiedenbeck, and D. Rhede, Self-diffusion of oxygen in mullite. *Journal of The European Ceramic Society*, **21**(14): p. 2577-2582 (2001).

19. F. Lofaj, S.M. Wiederhorn, F. Dorcakova, and K.J. Hoffmann. The effect of glass composition on creep damage development in silicon nitride ceramics. in *11th International Conference on Fracture*. Turin (2005).

20. K.J. Yoon, S.M. Wiederhorn, and W.E. Luecke, Comparison of tensile and compressive creep behavior in silicon nitride. *Journal of the American Ceramic Society*, **83**(8): p. 2017-2022 (2000).

21. G. Urbain, Y. Bottinga, and P. Richet, Viscosity of liquid silica, silicates and alumino-silicates. *Geochimica Et Cosmochimica Acta*, **46**: p. 1061-1072 (1982).

MICROSTRUCTURAL EVOLUTION OF CVD AMORPHOUS B-C CERAMICS HEAT TREATED: EXPERIMENTAL CHARACTERIZATION AND ATOMISTIC SIMULATION.

Camille Pallier, Georges Chollon, Patrick Weisbecker, Jean-Marc Leyssale, F. Teyssandier
CNRS-LCTS, Pessac France

Laboratoire des composites thermostructuraux, Université de Bordeaux 1, Pessac, France.
1 Chimie de la Matière Condensée de Paris, Collège de France, Paris, France.

ABSTRACT:
B-C materials are used to heal the cracks in ceramic matrix composites (CMC). B-C is deposited by CVD from BCl_3-CH_4-H_2 mixtures at high temperature, under reduced total pressure. Whatever their composition (B/$C_{at.}$ = 1.73-2.46), the as-deposited materials are nearly amorphous and consist of a common very disordered boron carbide phase (B_xC). When submitted to high temperatures, these materials undergo microstructural transformations, which are studied in this paper. The amorphous B-C phase is gradually transformed into free aromatic carbon and B_4C nanocrystals, in agreement with thermodynamics. This peculiar crystallization process is expected to lead to a significant time/temperature-dependent change of density and mechanical properties. The structure and crystallization behavior in inert atmosphere of the B-C ceramics were investigated as a function of the annealing temperature and time. Ex situ analyses were conducted by heat-treating the specimens under high vacuum at different temperatures/durations. The structure was characterized by Raman microspectroscopy, X-ray diffraction and transmission electron microscopy. The structural model was confirmed by liquid quench molecular dynamics simulation. High temperature tensile tests were performed on model 1D composites consisting of B-C coatings deposited on soft carbon monofilaments. These micro tensile tests allow the evaluation of the changes of (i) the volume (or density), (ii) the Young's modulus, (iii) the creep rate and (iv) the thermal expansion of the coatings. The results are discussed on the basis of elemental composition, initial structure and processing conditions.

KEYWORDS: Chemical Vapor Deposition (CVD), B-C Ceramics, Boron Carbide, Structure, Crystallization, Amorphous Materials, Short-range Ordering

INTRODUCTION

Boron carbide (B_4C) offers many attractive properties: high melting point and hardness (among the hardest materials), high modulus of elasticity, low specific weight, high chemical stability, large neutron capture section and good high-temperature thermoelectric properties... It is currently used for applications such as light weight armors, grinding, cutting tools, ceramic bearing… In the present work, boron carbide is used as a healing phase in ceramic matrix composites (CMC)[1,2]. These composites are composed of a preform made of woven carbon or ceramic fibers that confers mechanical resistance to the part. The preform is embedded in a ceramic matrix to produce a dense material and protect the fibers from oxidation. CMC are used in aeronautic engines at high temperature and in oxidative atmospheres. Their lifetime is mainly dependent on the oxidation of the reinforcing fibers. Lifetime can be improved by protecting the fibers from atmospheric oxygen or water vapor. These species diffuse towards the fibers through the cracks of the matrix. Oxidation of boron carbide takes place at rather low temperature to form a glass that fills the matrix cracks by capillarity and delays the diffusion of oxidative species, thus improving very much the material lifetime.

Boron carbide can be synthesized by means of numerous processes. For a comprehensive review, the reader is referred to the paper of Suri et al.[3]. Boron carbide can be grown from a gas phase by means of the chemical vapor deposition (CVD) process. A wide variety of boron or carbon bearing species can be used for that purpose such as halides, borane or hydrocarbons. When boron halides are used, hydrogen is required as a reducing agent. The B-C phase used to protect CMC from oxidation is synthesized by chemical vapor deposition (CVD). The as-deposited B-C phase may undergo structural rearrangement or crystallization when submitted during a long period of time at temperatures higher than their elaboration temperature. These structural rearrangements, which are induced by temperature activated processes, are the topic of the present paper.

Structure of Crystalline Boron Carbide

The structure of boron carbide has been a matter of debate for a long time. The recent review paper from Domnich et al.[4] presents the various controversies about its structure and states the up to date understanding of this complex compound, furthermore presenting a wide range of composition. B_4C has a rhombohedral lattice structure composed of B-rich icosahedra ($B_{12-X}C_X$) linked together by means of 3-atom linear chains disposed along the (111) rhombohedral axis (figure 1). These icosahedra reflect the ability of covalent boron compounds to form caged structures. The 12 atoms icosahedron structure includes i) six polar sites composed of the six atoms that form the top and bottom triangular faces and ii) six equatorial sites forming an hexagonal chair. The equatorial sites are those to which the 3-atom linear chains are bonded and the polar sites are directly linked to atoms in polar sites of neighboring icosahedra by covalent bonds. It is now established from theoretical simulation of NMR spectra based on density functional theory (DFT) (Mauri et al.[5]) that the pattern of the 3-atom chain is composed of a CBC sequence, the two carbon atoms being bonded to boron atoms on equatorial sites of adjacent icosahedra. At the B_4C stoichiometric composition (i.e. $B_{12}C_3$ or 20 %at C), the unit cell is composed of a $B_{11}C$ icosahedron with the carbon atom located on a polar site and a CBC chain. $B_{10}C_2$ icosahedra can be present as defects in some B_4C samples, but do not exceed a few percents. In these icosahedra, the two carbon atoms are found to be in antipodal polar positions.

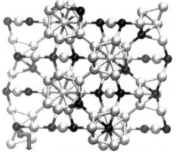

Figure 1: B_4C structure composed of $B_{11}C$ icosahedra and CBC chains: boron atoms in white and carbon atoms in black.

EXPERIMENTAL

B-C materials were deposited in a hot wall reactor operating at a temperature ranging from 850 to 1050°C under reduced pressure (2-12 kPa). A BCl_3-CH_4-H_2 mixture was used for that purpose[6,7]. Deposits were carried out either on flat SiC substrates (SiC polycrystalline wafer) for microanalysis and microstructural characterizations, or on XN05 carbon fibers (isotropic pitch-based), to study the volume variations induced by heat treatments. Due to their low modulus (60 GPa), these fibers do not significantly alter the deformations of the B-C coating.

The morphology of the coatings was characterized by scanning electron microscopy (SEM). The chemical composition was measured either by electron probe microanalysis (EPMA) carried out on the flat specimens, when the thickness of the coatings was sufficiently thick, or by in-depth Auger electron spectroscopy profiling, for thinner coatings. High purity CVD-SiC and sintered B_4C were used as the Si, C and B standards. The structure at long range was studied by X-ray diffraction, functioning either in the Bragg-Brentano geometry mode for the powder specimens, or parallel beam / glancing angle, for coatings on C fiber tows. Transmission electron microscopy was used to identify the crystalline phases in the coatings at a high spatial resolution. The structure at intermediate range was evaluated, through vibration properties, by Raman microspectroscopy. The lateral resolution is approximately 1 μm and the thickness probed varies from a few tenths to several hundreds of nanometers, depending on the composition and the structure.

Mechanical testing procedure

Tensile tests were performed from room temperature up to 1300°C under secondary vacuum (<10−3 Pa) on microcomposites with a gauge length of 50 mm. The device used for that purpose is detailed elsewhere[8]. Samples are held between two graphite grips using a carbon-based cement. Specimen heating is obtained by the application of an electric current, which, in the case of a single carbon fiber, leads to a uniform temperature being generated along the specimen. Tests were performed by applying and maintaining constant during sample testing a very low stress of 20 MPa onto the microcomposites. The displacement which had to be applied to maintain the imposed stress thus balanced the longitudinal expansion or contraction of the sample induced by the heat-treatment.

Composition and structure of the as-Deposited Material

The as deposited boron carbide material includes an excess of carbon with respect to the stoichiometry of crystalline B_4C (B/C_{at}=2.46). The equilibrium state should accordingly belong to the two-phase domain: B_4C + graphitic carbon. The transmission electron microscopy observation of the material in high-resolution mode (HR-TEM) does not reveal lattice fringe of any type. The selected area electron diffraction (SAED) pattern shows a limited number of complete and diffuse rings, which are typical features of an isotropic and amorphous material. The X-ray diffraction pattern (not shown) and Raman spectrum (Figure 2) of the material confirm the absence of pure carbon phases (sp^2 type, graphite) or crystalline boron carbide.

The Raman spectrum of the as-deposited material is composed of two very broad bands around 400-700 cm^{-1} and 850-1350 cm^{-1}, which are typical of an almost amorphous state. The band at 850-1350 cm^{-1} can be attributed to breathing modes of icosahedron-like units by comparison with the spectrum of crystalline B_4C ($c\text{-}B_4C$)[9]. A structural organization containing features close to the icosahedral units found in $c\text{-}B_4C$ should thus exist in the amorphous boron carbide ceramic ($a\text{-}B_xC$). Though free carbon is expected to be present in the deposit because of its atomic composition, the well-known D (1350 cm^{-1}) and G (1580 cm^{-1}) bands of graphitic carbons are strikingly missing. Accordingly, instead of forming sp^2 C-C bonds, the carbon atoms in excess might all be bonded to boron atoms inside the amorphous material.

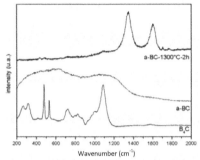

Figure 2 : Raman spectra of a-BC material: as-deposited (a-B$_x$C) and heat-treated during two hours at 1300°C (a-BC-1300). The spectrum of crystalline B$_4$C is also shown for comparison.

Computer Simulation

An amorphous model containing 216 atoms (154 boron and 62 carbon) at the density of the as-prepared ceramic (2.47 g/cm^3), was produced using Car Parrinello molecular dynamic (CPMD) simulations[10]. To do so we have simulated, at the atomic scale, the rapid quench of the system from 4000 to 0K. Initially in the liquid state, the system has frozen around 2000 K to form a disordered glass. A snapshot of the resulting a-BxC model, as obtained at the end of the quench is shown in Figure 3. These simulations were validated by comparing the atomic structure factor S(Q) and the reduced pair distribution function G(r) computed from the model with those obtained by neutron diffraction carried out on as-deposited CVD material of the same composition. Every peak of the experimental functions is very well reproduced by the model, both in terms of location and intensity. This enabled us to assume that the nano-structural model is a realistic representation of the experimental material.

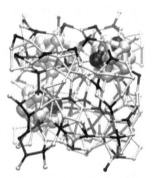

Figure 3: Snapshot of the atomistic a-B$_{154}$C$_{62}$ model showing three icosahedra (B$_{12}$, B$_{11}$C and B$_{10}$C$_2$) in an amorphous matrix (white: boron atoms, black: carbon atoms; large spheres and bonds: icosahedral units).

As can be seen in Figure 3, the model includes three 12-atoms icosahedral units. As in c-B$_4$C, these icosahedra have an excess of boron atoms with respect to the material's stoichiometry with one

$B_{11}C$ unit, dominating in the crystal and a B_{12} and a $B_{10}C_2$ units, the main substitutions of B and C atoms in c-B_4C^4. In the amorphous model, these icosahedral units are embedded in a disordered matrix composed of incomplete icosahedra, still mainly formed by six-fold atoms, and of areas of lower density showing lower coordination numbers.

Among all atoms, 66 boron atoms (42.9 %) and 3 carbon atoms (4.8 %) have the sixfold coordination found in icosahedral units. These atoms form many three-center bonds lying almost exactly in the middle of a three atoms triangle.

The occurrences of all the boron and carbon neighborhoods were also calculated. Among the 66 hexa-coordinated boron atoms (43 % of boron atoms of the model), 13, 31 and 17 respectively show B_6, B_5C and B_4C_2 environments that are found in c-B_4C in similar proportions. Only three carbon atoms (those in the icosahedra) and are found in B_6 environments. A majority of carbon atoms (61.3 %) are fourfold carbon atoms (sp^3 hybridization) as for the carbon atoms located at the two ends of the inter-icosahedral CBC linear chains in c-B_4C. However, not a single twofold boron atom, characteristic of the CBC chains, could be detected. Instead, 23 (15 %) threefold (sp^2) boron atoms are found (in mainly C_3 (8) and C_2B (11) environments). Also, among the 62 carbon atoms of the model, only three are threefold, indicating the absence of free graphitic carbon in the model (six threefold C atoms being necessary to form a single aromatic ring). A model of the amorphous material can thus be drawn on the basis of 12 atoms icosahedral units in an amorphous matrix made of sp^2 boron atoms and sp^3 carbon atoms.

Strain Variation of the Microcomposite at High Temperature Under Very Low Stress

The tests were carried out on 4.4 µm thick B-C coatings deposited on XN05 fibers (diameter 10 µm). The composition of the material deposited on the carbon fibers contains more carbon in excess than the material deposited on flat SiC substrates. The microcomposites hence obtained had a matrix volume fraction of 0.72. The surface of the deposit was extremely smooth and did not reveal any flaw. The SEM observation of a cross-section did not reveal either any variation of morphology throughout the thickness of the deposit (figure 4). Transmission electron microscopy image of the as deposited a-$B1.73C$ material in high-resolution mode (HR-TEM), did not reveal any lattice fringe. The selected area electron diffraction (SAED) pattern showed a limited number of complete and diffuse rings, which are typical features of an isotropic and amorphous material. In spite of their difference in composition, the materials deposited on SiC or carbon fiber are quite similar and both amorphous.

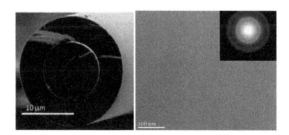

Figure 4: High resolution bright field transmission electron microscopy image of the as-deposited a-$B1.73C$ material. The corresponding selected area electron diffraction pattern is shown in the inset.

Tensile elongation measurements under very low stress (20 MPa) were carried out on XN05/$B_{1.73}C$ microcomposites at temperatures higher than the deposition temperature (900°C). These tests were aimed at observing the influence of the structural rearrangement on the density variation of the material. The elastic modulus of the deposit has been determined using the rule of mixture by load cycling tensile tests carried out at room temperature. The value (253 ± 21 GPa) for the as deposited material is much lower than that of polycristalline B_4C: 450 GPa.

Four testing temperatures were studied: 1000, 1100, 1200 and 1300°C. The deformation behavior of the microcomposite, either elongation (positive) or shrinkage (negative), is plotted versus time in figure 5. The tests revealed very different behaviors according to temperature.

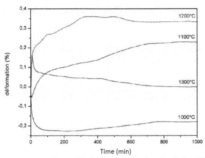

Figure 5: Elongation measurements under a 20 MPa tensile load of the $XN05/B_{1.73}C$ microcomposite between 1000 and 1300 °C. The transient domain corresponding to the initial elastic loading and heating of the specimen is not represented.

The observed deformations strongly vary according to time and temperature.
- At 1000 °C, a fast shrinkage stage occurs. Almost no deformation is observed between 100 and 300 min and a slight elongation is observed for longer times.
- At 1100 °C, a fast shrinkage is still observed for a very short period of time (6 min) and followed by an expansion regime with a strain rate decreasing with time.
- At 1200 °C, no shrinkage is observed. Only a fast expansion occurs at first which tends to diminish for long durations. Some erratic behavior is observed after 300 min.
- At 1300 °C, a strong and short elongation domain immediately followed by a short and strong shrinkage is observed at the beginning of the test. The shrinkage further gently decreases. No deformation is observed after 800 min.

It should be noticed that the observed deformations are much higher than those measured on the fiber alone submitted to an equivalent load. These variations are therefore features typical of the behavior of the ceramic matrix deposited on the fiber (which is stable in these conditions). In order to improve the understanding of these variations, post mortem characterizations were carried out by SEM and Raman analyses on cross-sections of the microcomposites as well as by TEM. SEM observations of the microcomposite heat-treated at 1000°C did not reveal any morphology change and only very small amounts of carbon are revealed on the Raman spectra. The morphology of microcomposites heat-treated at 1100°C is unchanged as revealed by SEM. TEM bright field images show the formation of small (a few nm) B_4C grains in the vicinity of the fiber/matrix interface, while the rest of the matrix remains apparently amorphous. However, the comparison between the electron diffraction patterns of the amorphous part of the matrix before and after heat-treatment is indicative of a short-range structural variation after 1000 min (figure 6).

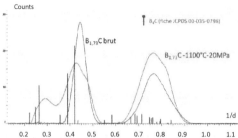

Figure 6: selected area electron diffraction patterns of the amorphous part of B1.73C before and after heat-treatment at 1100°C.

The morphology of the B-C materials heat-treated at 1200 and 1300°C is markedly different. SEM observations (figure 7) of the cross-sections of the deposit show a granular morphology, suggesting a crystallization of the matrix. The D and G peaks of the Raman spectra have increased but the presence of B_4C is not clearly revealed (the weak B_4C peaks being possibly hidden by the carbon peaks). The smoother morphology observed between 1200 and 1300°C is probably due to a decrease of the intergranular residual porosity, as observed during the sintering of a porous green body.

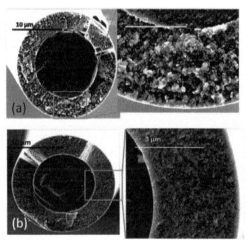

Figure 7: SEM images MEB of microcomposites heat-treated under 20 MPa at (a) 1200 °C et (b) 1300 °C

DISCUSSION

The shrinkage observed at 1000 and 1100°C during short period of times may be associated to the local structural rearrangement of the amorphous B-C material. Indeed, even for very long heat-treatments, free carbon is not formed or extremely slowly. Such a densification may result from the rearrangement of tricoordinated or hexacoordinated boron sites ($\underline{B}C_3$ et $\underline{B}B_{6-x}C_x$) and tetracoordinated carbon sites ($\underline{C}B_4$) to form more regular icosahedral structures, a preliminary step to the crystallization of B_4C. This phenomenon takes only about 6 minutes at 1100°C and is followed by the swelling of the material as the result of carbon segregation. The excess of carbon that is present in the amorphous material as Csp3 is thus transformed in disordered Csp2 inducing volume expansion. Indeed, it is well known that transformation of diamond into graphite induces a 60% volume expansion.

At 1200 °C, only a fast volume expansion resulting from the segregation of free carbon is observed. Though crystallized B_4C is not observed by Raman spectroscopy, the morphology of the matrix (as observed by SEM) is indicative of a crystallized material.

The transformations occurring at 1300 °C are complex and may result from two competing phenomena: a swelling, for a short period of time induced by the rapid formation of Csp2 and a shrinkage, induced by the structural reorganization of both B_4C and free carbon. The reduction of the porosity and the rearrangement these two phase at a micro-scale is probably the main driving force of the long term shrinkage. The amplitude of the shrinkage is correlated to the amount of free carbon generated in the preceding carbon segregation.

CONCLUSION

Quasi-amorphous boron carbide matrix was deposited by CVD at 900°C from a BCl_3-CH_4-H_2 gas mixture. Based on structural characterizations of the material and an atomistic modeling we can assume that the amorphous phase includes highly disordered icosahedron-like units (including $\underline{B}B_{6-x}C_x$ sites). The intericosahedral domain includes trigonal B-C3 environments, which are normally absent in crystalline boron carbide. It also includes carbon tetracoordinated sites (CB4), similar to the coordination of carbon atoms belonging to the CBC chain (though none of these chains were detected). The singular B local environment probably explains the absence of sp2 carbon, contrary to what would be expected in a material of the same composition at equilibrium.

Both structural analyses and deformation tests were used to understand the structural rearrangements at various scales induced by heat treatments at various temperatures (ranging from 1000 to 1300°C).When the temperature increases, the first step is a short range rearrangement that results in the formation icosahedron units (i.e. by completing the already present disordered icosahedron-like units), as well as a strong decrease of the proportion of trigonal B-C3 sites, in favor of intra-icosahedral sites of rhombohedral B_4C. The material becomes more dense (shrinkage). The second step is the carbon segregation and formation of sp2 C-C3 (hexagonal) sites, which are initially absent in the material. This leads to a volume expansion of the material. The last step is the complete crystallization of the B_4C phase accompanied by the formation of inter-icosahedral C-B-C chains and, which can take place only above 1200°C whatever the duration of the heat treatment.

REFERENCES

[1] F. Lamouroux, S. Bertrand, R. Pailler, R. Naslain, M. Cataldi, Oxidation-resistant carbon-fiber-reinforced ceramic-matrix composites,*Comp. Sci. Technol.*, **59**, 1073-1085 (1999)

[2] L. Vandenbulcke, G. Fantozzi, S. Goujard, M. Bourgeon, Outstanding ceramic matrix composites for high temperature applications, *Adv. Eng. Mater.*, **7**, 137-142 (2005)

[3] A. K. Suri, C. Subramanian, J. K. Sonber, T. S. R. Ch. Murthy, Synthesis and consolidation of boron carbide: a review, *International Materials Review,* **55**, 1, 4-40 (2010)

[4] V. Domnich, S. Reynaud, R.A. Haber, M. Chhowalla, Boron Carbide: Structure, Properties, and Stability under Stress, *J. Am. Ceram. Soc.*, **94**, 3605-3628 (2011)

[5] F. Mauri, N. Vast, C. Pickard, *J. Phys. Rev. Lett.*, **87**, 085506 (2001)

[6] J. Berjonneau, G. Chollon, F. Langlais, Deposition Process of Amorphous Boron Carbide from $CH_4/BCl_3/H_2$ Precursor *J. Electrochem. Soc.*, **153**, 12, C795-C800 (2006)

[7] J. Berjonneau, F. Langlais, G. Chollon, Understanding the CVD process of (Si)–B–C ceramics through FTIR spectroscopy gas phase analysis, *Surf. Coat. Technol.*, **201**, 7273-7285 (2007)

[8] C. Sauder, J. Lamon, R. Pailler, Thermomechanical properties of carbon fibres at high temperatures (up to 2000 C), *Composites Science and Technology*, **62**, 499–5(2002)

[9] D.R. Tallant, T.L. Aselage, A.N. Campbell, D. Emin, Boron carbide structure by Raman spectroscopy, *Phys Rev B*, **40,** 8, 5649-5656 (1989)

[10] C. Pallier, J.-M. Leyssale, A. T. Bui, P. Weisbecker, H. E. Fischer, C. Gervais, L. Truflandier, F. Sirotti, F. Teyssandier, G. Chollon, *Chemistry of materials to be submitted.*

DENSIFICATION OF SiC WITH AlN- Nd_2O_3 SINTERING ADDITIVES

Wu Laner, Jiang Yong, Sun Wenzhou, Chen Yuhong, Huang Zhenkun
School of MSE, Beifang University of Nationalities, Yinchuan, Ningxia, China

ABSTRACT

The present research concerns SiC ceramics sintered with AlN-Nd_2O_3. The sintered samples were analyzed by X-Ray diffraction, Electron Probe Microscopy and Scanning Electron Microscopy. The densities and mechanical properties of the samples such as Hardness and fracture toughness are also measured. Base on the experimental results, the relative density of the sintered body was above 96% with hardness of 18GPa(Hv). The microstructure of the samples shows compact but inequality of grains size. A solid solution of $Nd2Al_{1-x}Si_xO_3N_{1-x}C_x$ (x=0-0.5) can be observed in the SiC-AlN-Nd_2O_3 system.

INTRODUCTION

SiC is an excellent structural ceramic. Due to its strong covalent character and poor high temperature diffusion coefficient, it is very difficult to sinter. Rare earth oxides (R_2O_3, R could be Ce, Pr, Nd, Sm, Eu) and AlN together can be used as sintering additives. The AlN-R_2O_3 additive system can improve the densification of SiC ceramics. Solid solution of R_2AlO_3N could be formed in the SiC-AlN-R_2O_3 systems. The AlN-R_2O_3 system has been used as an effective sintering additive of SiC [1-4]. However, it is still not very clear about their high temperature reaction and the ternary phase relationship. The present research is focused on SiC ceramics sintered with AlN-Nd_2O_3 as sintering additives.

EXPERIMENTAL

The starting powders of the experiments used were α-SiC with 1.1 % O_2 (UF-15-A, H.C.Starck), AlN with 0.9 % O_2 (M11, H.C.Starck), Nd_2O_3 (Baotou Research Institute of Rare Earth, purity > 99.9%). The rare earth oxides were calcined at 1200°C for 2 hours to remove the water before being used. Selected compositions were prepared as SiC: AlN: Nd_2O_3 = 2:2:1(molar ratio). The powder mixtures were ground in an agate mortar for 2 hours with alcohol. After drying, the prepared powder was put into a round graphite mold of 10mm diameter and hot pressed. To prevent the sample from sticking to the mold, the latter was lined with BN. The hot pressing conditions used were: Ar atmosphere under 30 MPa pressure at 1500°C for 2 hours. After releasing the pressure, the temperature was held for 1-3 hours. The densities of the sintered body were measured based on Archimedes Law. Hardness of the sample was tested by microhardness tester (HXS-1000, Shanghai, China; 3Kg); Fracture toughness were calculated by the crack lengths of the impression. Microstructure of the sample was observed by scanning electronic microscope (SHIMADZU, SSX-550,Japan); The phase composition of the sintered samples were analyzed by X-ray diffractometer (SHIMADZU XRD-6000) with CuKα ray in 0.2 degrees scan step, 2 degrees / minute. The element compositions of the samples were analyzed by using electron probe micro analyzer JXA−8100 (EPMA), JEOL, with beam diameter of 0.5μm and a beam current of 1×10^{-8}A. The surfaces of the samples were coated with carbon before the analysis was carried out.

RESULTS AND DISCUSSION

1. Properties and microstructure of the sintered body

The properties of the sintered body were as follow: The relative density of the sintered body is above 96% and the Vickers hardness is 18GPa. The results are not quite satisfactory. The reason for that might be the lower sintering temperature.

Microstructure of the sintered sample is shown in Fig 1, and indicates that the unequal grains size varied from submicron to dozens of micron. It can be determined from back scatting image contrast of the picture that the white bigger grains are Nd2O3, while black grains are AlN and SiC. The non-uniform microstructure might cause by aggregation of starting powders. It should be avoided by better dispersion and finer particles of starting powder in further research.

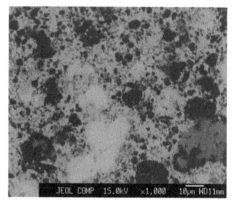

Fig. 1 Microstructure of polished surface of SiC-AlN-Nd $_2$O$_3$ sintered sample at 1500°C hot pressing for 2 hours.

2. Phase analysis of AlN-SiC- Nd$_2$O$_3$ system

The experimental system of SiC-AlN-Nd2O3 has molar ratio of SiC:AlN:Nd$_2$O$_3$ = 2:2:1. The XRD pattern of the phase composition of the sample is shown in Fig 2. It can be seen in Fig 2 that besides SiC , AlN and Nd$_2$O$_3$, there is a new phase of Nd$_2$AlO$_3$N in the system. The phenomenon has been reported in our previous research [26]. As we know, in the common liquid-phase sintering of SiC ceramic, the oxide impurities in the starting powders of SiC and AlN normally dissolve into liquid phase. In the solid-phase sintering stage, however, these impurities participate in the solid-state reaction with other components forming some salt compound as a small subordination phase. In present systems of AlN-SiC- Nd$_2$O$_3$, the impurities of Al$_2$O$_3$ from the AlN starting powder and SiO$_2$ from SiC powder reacted with Nd$_2$O$_3$ forming a small amount of salt compounds. These salt compounds could be formed by impurity Al$_2$O$_3$ and SiO$_2$ directly reacted with Nd$_2$O$_3$ as follows: $Al_2O_3 + Nd_2O_3 \rightarrow Nd_2O_3 \cdot Al_2O_3 \rightarrow NdAlO_3$ and $SiO_2 + AlN + Nd_2O_3 \rightarrow$ NdAlO$_3$ + NdSiNO$_2$ (K-phase).

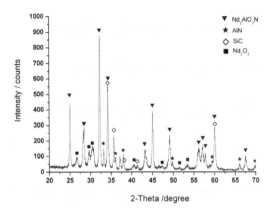

Fig 2. XRD of SiC:AlN:Nd$_2$O$_3$ = 2:2:1 composition HPed 1550C in Ar.

3. Phase relations in AlN-SiC-Nd$_2$O$_3$ system

The phase relationship of the AlN-Nd$_2$O$_3$ binary system has been published by one of the authors [11]. Within this system, the compound Nd$_2$AlO$_3$N, with molecular ratio of 1:1, tetragonal structure, and congruent melting point at 1750° C, was reported. The present research proved this compounds existence again. The EPMA analysis results of the Nd$_2$AlO$_3$N (1:1) phase in the 2:2:1 sample are shown in Table 1. It can be seen from Table 1 that Al-N in Nd$_2$AlO$_3$N could be partially substituted by Si-C to form Nd$_2$Al$_{1-x}$Si$_x$O$_3$N$_{1-x}$C$_x$ (x=0-0.5) solid solution. Figure 3 shows subsolidus phase diagram of the AlN-SiC-Nd$_2$O$_3$ ternary system [12]. Both SiC and AlN have a same hexagonal wurtzite structure, and the bond lengths of SiC (1.89Å) and AlN (1.87 Å) are very closer, which provide them ability to form a solid solution. The phase relationship of the AlN-SiC system has been published by Zangvil and R. Ruh [13]. Within the system a continuous solid solution δ(2H) of (SiC)$_{1-x}$(AlN)$_x$ formed when the temperature was over 2000°C. A wide immiscible region of two phases δ1 and δ2 existed while the temperature was below 2000° C [13]. The synthesis of δ1 or δ2 solid-solution, i.e. (SiC)$_{1-x}$(AlN)$_x$ or (AlN)$_{1-x}$(SiC)$_x$, at a much lower temperatures has been reported [14-17] .

Table 1   Cation ratios for the Nd$_2$Al$_{1-x}$Si$_x$O$_3$N$_{1-x}$C$_x$ phase in AlN:SiC:Nd$_2$O$_3$=2:2:1 sample by EPMA

Point	Nd	Si	Al	(Al+Si)/Nd	Si/(Al+Si)
1	24.86	5.37	7.60	0.52	0.41
2	27.38	6.33	7.45	0.47	0.50
3	29.11	6.11	6.14	0.42	0.50
4	24.36	6.56	6.82	0.55	0.49

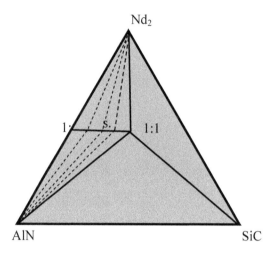

Fig. 3 Subsolidus phase diagram of AlN-SiC-Nd$_2$O$_3$ system

CONCLUSIONS

1. Liquid phase sintered SiC -AlN- Nd$_2$O$_3$ made through hot pressing of 1500°C resulted in a relative density above 96% and a Vickers hardness 18GPa.

2. The microstructure of the sample shows that the unequal grain size varied from submicron to dozens of micron. It might be improved by better dispersion and finer particles of the starting powder.

3. In this system, the existence of Nd$_2$AlO$_3$N (1:1) compound was verified. It was found that the Al-N could be partially substituted by Si-C forming a Nd$_2$Al$_{1-x}$Si$_x$O$_3$N$_{1-x}$C$_x$ (x: 0~0.5) solid solution. This solid solution was compatible with SiC.

ACKNOWLEDGEMENT

Financial support from the National Natural Science Foundation of China, NSFC50962001is gratefully acknowledged.

REFERENCES

[1] G. Magnani, F. Antolini, L. Beaulardi, E. Burresi, A. Coglitore, C. Mingazzini, "Sintering, high temperature strength and oxidation resistance of liquid-phase pressureless-sintered SiC–AlN ceramics with addition of rare-earth oxides", J. Euro. Ceram. Soc., 29 [11] 2411-2417 (2009).

[2] B. V. Manoj, M. H. Roh, Y. W. Kim, W. Kim, and S. W. Park, "Mechanical properties of SiC ceramics sintered with Re$_2$O$_3$(Re: Sc, Lu, Y) and AlN additives", Met. Mater. Int., 16 [2] 229-233 (2010).

[3] R.M. Balestra, S. Ribeiro, S.P. Taguchi, F.V. Motta, C. Bormio-Nunes "Wetting behaviour of Y$_2$O$_3$/AlN additive on SiC ceramics".
J. Euro. Ceram. Soc., 2006,26(16) 3881–3886

[4] Wu Laner, Chen Yuhong, Jiang Yong, Huang Zhenkun. "Liquid sintering of SiC with AlN–Re$_2$O$_3$ Additives". J.of the Chinese ceramic society 2008, 36，[5] 593-596

[5] E. Parthe (Ed.) "Crystal Chemistry of Tetrahedral Structures,"pp. 25-27, Gordon, New York　(1964).

[6] G.A. Jeffrey, G.S. Parry, and R.L. Mozzi, "Study of the Wurtzite-Type Binary Compounds. I. Structures of Aluminum Nitride and Beryllium Oxide". J. Chem. Phys., 25(5), 1024　　(1956).

[7] T.S. Yan，C.K. Kuo, W.L. Han, Y.H. Qui, and Y.Z. Huang, "Phase relationship in the series of the BeO-R$_2$O$_3$ systems (R=La,Nd,Sm,Gd,Dy,Ho,Er,Yb,Lu, and Y)" J. Amer. Ceram. Soc., 66[12], 860 (1983).

[8] R. Marchand, "Oxynitrides with potassium nickel (II) tetrafluoride structure. Ln$_2$AlO$_3$N compounds (Ln = lanthanum, neodymium, samarium)".　　C. R. Acad. Sci. Ser. C 282 (7), 329-31 (1976).

[9] Z. K. Huang, P. Greil, and G. Petzow, "Formation of α-solid solution in the system Si$_3$N$_4$-AlN-Y$_2$O$_3$ ",　J. Am. Ceram. Soc., 66 [6] C96-C97 (1983),

[10] Z. K. Huang, T. Y. Tien, and T. S. Yen, "Subsolidus Phase Relationships in Si$_3$N$_4$-AlN-Rare-Earth Oxide Systems",　J. Am. Ceram. Soc., , C69 [10] C241-242 (1986),

[11] Z. K. Huang and D.S. Yan (T.S. Yen), "Compound formation and melting behavior in the AB compound and rare earth oxide systems" J. Solid State Chemistry, 85, 51-55 (1990)

[12] Yuhong Chen, at el. "Phase Relations in SiC-AlN-R$_2$O$_3$ (R=Nd, Gd, Yb, Y) Systems" Journal of Phase Equilibria and Diffusion.　(in publication)

[13] Zangvil and R. Ruh, "Phase Relationships in the Silicon Carbide– Aluminum Nitride System," J. Am. Ceram. Soc., 82 [9], 2481–2489　(Sep. 1999).

[14] Chen Kexin, Jin Haibo, Zhou Heping and José M.F. Ferreira,"Combustion synthesis of AlN–SiC solid solution particles", JECS, Volume 20, Issues 14–15, Pages 2585-2590 (December 2000)

[15] Pezoldt, , R.A. Yankova, A. Mücklich, W. Fukarek, M. Voelskow, H. Reuther   and W. Skorupa, "A novel (SiC)$_{1-x}$(AlN)$_x$ compound synthesized using ion   beams". Nuclear Instruments and Methods in Physics Research Section B: Beam Interactions with Materials and Atoms, Volume 147, Issues 1-4, Pages 273-278 (1999),.

[16] R. Roucka, J. Tolle, A. V. G. Chizmeshya, P. A. Crozier, C. D. Poweleit, D. J. Smith, I. S. T. Tsong, and J. Kouvetakis, "Low-Temperature Epitaxial Growth of the Quaternary Wide Band Gap Semiconductor SiCAlN", Phys. Rev. Lett. 88, 206102 (2002)

[17] R.-C. Juang, C.-C. Chen, J.-C. Kuo, T.-Y. Huang, Y.-Y. Li, "Combustion synthesis of hexagonal AlN-SiC solid solution under low nitrogen pressure".Journal of Alloys and Compounds, Volume 480, Issue 2, , Pages 928-933 (8 July 2009)

SOLID-SOLUTION OF NITROGEN-CONTAINING RARE EARTH ALUMINATES R_2AlO_3N (R=Nd and Sm)

Yong Jiang, Laner Wu, Zhengkun Huang
School of MSE, Beifang University of Nationalities, Wenchang Road, Xixia District, 750021
Yinchuan, Ningxia, China

ABSTRACT

In the $SiC-AlN-R_2O_3$ (R=Nd, Sm) systems by Si-C partially substituted for Al-N in the R_2AlO_3N compound (R_2O_3:AlN=1:1) the formation of solid solution with formula $R_2Al_{1-x}Si_xO_3N_{1-x}C_x$ (for R=Nd, $0 \leq x \leq 0.5$; for R=Sm $0 \leq x \leq 0.4$) was first found. Impurities of Al_2O_3 from AlN and SiO_2 from SiC starting powders often caused the formation of some salt compounds (RAP type). Mechanism of forming salt compounds by the reactions of Al_2O_3 and SiO_2 impurities with rare earth oxides was discussed.

INTRODUCTION

As well known, SiC is an excellent structural ceramic. But due to its strong covalent character and poor high temperature diffusion coefficient, it is very difficult to be sintered. The $AlN-R_2O_3$ system has been used as an effective sintering additive of SiC for many years [1-4]. However, it is still not very clear about their high temperature reaction and the ternary phase relationship. A series of the phase relationships of $AlN-R_2O_3$ (R=La, Nd, Sm, Eu, Gd, Dy, Er, Yb and Y) systems have been published [5-9]. There existed a series of 1:1 compounds, R_2AlO_3N (R=Ce, Pr, Nd, Sm, Eu) [8, 9] in the light rare earth systems of $AlN-R_2O_3$. With the heavy rare earth oxides after Gd, the compound could not generate.

We have reported the phase relation of $SiC-AlN-Nd_2O_3$ system [10]. In binary Nd_2O_3 - AlN system only a compound Nd_2AlO_3N exists, which forms solid solution by Si-C partially substituted for Al-N.

Present work studies the formation of R_2AlO_3N (1:1) (R=Nd, Sm) solid solution and some salts caused by reaction of the impurities of Al_2O_3 and SiO_2 in starting powders with R_2O_3.

EXPERIMENTS

The starting powders of the experiments used were AlN with 0.9 % O_2 (M11, H.C. Starck), α-SiC with 1.1 % O_2 (UF-15-A, H. C. Starck), Nd_2O_3, Sm_2O_3 (Baotou Research Institute of Rare Earth, purity > 99.9%). The rare earth oxides were calcined at $1200^{\circ}C$ for 2 hours to remove the water content before being used. Some selected compositions, special $SiC:AlN:R_2O_3=2:2:1$, were prepared.

The powder mixtures were mixed and ground in an agate mortar for 2 hours with alcohol. After dried up, the prepared powder was put into a round graphite mold of 10mm diameter and hot pressed. To prevent the sample from sticking to the mold, latter was lined with BN. The conditions of hot pressing used were: in Ar atmosphere, under the pressure of 30 MPa, at 1500-$1600^{\circ}C$ hot pressing for 1-2 hours. The pressure-less sintering was also used to synthesis sample.

The phase compositions of the sintered samples were analyzed by SHIMADZU XRD-6000 X-ray diffraction (XRD) with CuKα ray in 0.2 degrees scan step, 2 degrees / minute. The element compositions of the samples were analyzed by using electro probe micro analyzer JXA -8100 (EPMA), JEOL, with beam diameter of 0.5μm, beam current of 1×10^{-8}A. The surfaces of the samples were coated with carbon before the analysis carried out.

RESULTS AND DISCUSSION

1. Formation of R_2AlO_3N (1:1) solid solution

The phase relationship of $AlN-Nd_2O_3$ binary system has been published by one of the authors [9] before. Within this system a compound Nd_2AlO_3N, with molecular ratio of 1:1, tetragonal K_2NiF_4 type structure [8], congruent melting point at 1750° C, was reported. Between $Nd_2AlO_3N-Nd_2O_3$ located a lowest eutectic point at 1660° C [9]. The present research proved this compound existence again through solid-state reaction via hot-press of the composition at 1600 ° C. The EPMA analysis results of the Nd_2AlO_3N (1:1) phase in the 2:2:1 sample are shown in Table 1. It can be found from Table 1 that Al-N in Nd_2AlO_3N could be partially substituted by Si-C to form $Nd_2Al_{1-x}Si_xO_3N_{1-x}C_x$ ($0 \leq x \leq 0.5$) solid solution. The composition of vertex point in the triangle $AlN-SiC-Nd_2AlO_3N$ (ss) has the largest solid solubility x=0.5 (see Table 1). Figure 1 shows the XRD patterns of the three-phase coexistence in the composition $AlN:SiC:Nd_2O_3=2:2:1$ (Fig. 1 a), where a small amount of Nd_2O_3 always remains although several tests repeated and holding sintering time prolonged. The reason is that some coarse grains of Nd_2O_3 powder used with grain size >5μm prevent themselves from reaching complete reaction [10].

Table 1 Cation ratios in the $R_2Al_{1-x}Si_xO_3N_{1-x}C_x$ phase for $AlN:SiC:R_2O_3 = 2:2:1$ sample by EPMA

Samples	R	Si	Al	(Si + Al)/R	X Si/(Al+Si)
AS-Nd	26.95	6.33	6.80	0.49	0.48 ($\approx$0.5)
AS-Sm	20.57	5.22	7.05	0.59	0.42 ($\approx$0.4)

* Average of three points.
** Data of small gas elements are too scattered to include.

The $AlN:SiC:Sm_2O_3 = 2:2:1$ composition has the similar phase relation of three phases coexistence of $AlN + SiC + Sm_2AlO_3N(ss)$ (Fig. 1 b), like that of Nd-composition. It forms also solid solution with the formula $Sm_2Al_{1-x}Si_xO_3N_{1-x}C_x$ ($0 \leq x \leq 0.4$, Table 1) by Si-C partially substituted for Al-N in Sm_2AlO_3N.

In addition, a few $RAlO_3$ (RAP type, R=Nd and Sm) was detected in some samples. It is because that a trace of Al_2O_3 impurity in AlN powder reacted with R_2O_3. Like $SiC-La_2O_3$ binary system [11], in Ar atmosphere there was no new phase formed for either $SiC-Nd_2O_3$ or $SiC-Sm_2O_3$ binary systems. The trend of salt compound formability with R_2O_3 is AlN > SiC with increasing their covalence.

The subsolidus phase diagram of the $AlN-SiC-Sm_2O_3$ ternary system shows in Figure 2, it likes that of the Nd-system. Just the tie-line of $SiC-Sm_2AlO_3N(ss)$ is shorter than that of $SiC-Nd_2AlO_3N(ss)$.

2. Formation of salt compounds RAP (R=Nd, Sm)

As we known that in the common liquid-phase sintering of SiC ceramic the oxide impurities in the powders of SiC and AlN normally dissolve into liquid phase. In the solid-phase sintering stage, however, these impurities participate in the solid-state reaction with other components forming some salt compound as small subordination phase. In present systems of AlN-SiC-R₂O₃ (R=Nd, Sm) the impurities Al₂O₃ from AlN and SiO₂ from SiC reacted with R₂O₃ forming a little amount of salt compounds. RAlO₃ was tested by XRD to have the structures of perovskite (RAP) type. The stability of salt compounds in R-SiAlON system related with rare-earth series shows in the selected Table 2 [12]. NdAP and SmAP are just stable in light rare earths range (see Table 2).

The salt compounds could be formed by impurity Al₂O₃ and SiO₂ directly reacted with R₂O₃ as following:

(1) $Al_2O_3 + R_2O_3 \rightarrow R_2O_3 \cdot Al_2O_3 = 2RAlO_3$ (RAP),

(2) $SiO_2 + AlN + R_2O_3 \rightarrow RAlO_3$ (RAP) + $RSiNO_2$ (K-phase),

No any RAP solid solution with SiO₂ was reported. The impurity SiO₂, however, still could form some nitrogen-containing salt compound like RSiNO₂ (K-phase), but it is too small to be detected in present work.

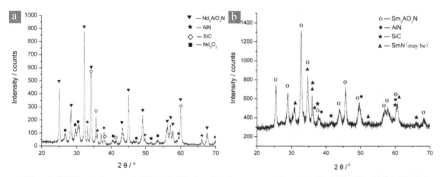

Fig. 1 XRD pattern of SiC:AlN:R₂O₃=2:2:1 composition (a) R=Nd HPed (b) R=Sm pressureless sintered

Table 2. Stability of some phases in R-SiAlON systems [12]

Phases*	La	Ce	Pr	Nd	Sm	Eu	Gd	Dy	Y	Er	Yb
RalO₃ (RAP)	+	+	+	+	+	+	+	+			
R₄Al₂O₉ (RAM)						+	+	+	+	+	+
R₃Al₅O₁₂ (RAG)							+	+	+	+	+
K-phase	+	+	+	+	+			+			
H-phase	+	+	+	+	+	+	+	+	+	+	+
M'-phase	+	+	+	+	+	+	+	+	+	+	+
J-phase	+	+	+	+	+	+	+	+	+	+	+
R₂AlO₃N	+	+	+	+	+						
R₂SiAlO₅N (B-phase)								+	+	+	+

* K-phase: $R_3Si_3O_6N_3$;   M'-phase: $R_2Si_{3-x}Al_xO_{3+x}N_{4-x}$;
   H-phase: $R_{10}(SiO_4)_6N_2$;   J-phase: $R_4Si_2O_7N_2$.

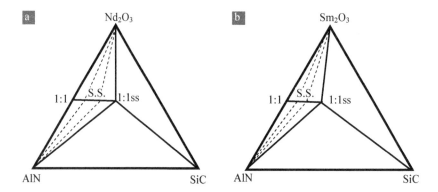

Fig 2. Tentative phase diagram of (a) AlN-SiC-Nd_2O_3 and (b) AlN-SiC-Sm_2O_3 systems

CONCLUSIONS
1. Phase relations of the AlN-SiC-R_2O_3 (R=Nd, Sm) systems were determined.
2. In AlN-R_2O_3 subsystem, R_2AlO_3N (1:1) solid solution with $R_2Al_{1-x}Si_xO_3N_{1-x}C_x$ (for R=Nd, $0 \leq x \leq 0.5$; for R=Sm, $0 \leq x \leq 0.4$) formula was first found by Si-C partially substituted for Al-N. This solid solution is compatible with SiC.
3. Mechanism of forming salt compounds by the reactions of Al_2O_3 and SiO_2 impurities in the powders with rare earth oxides was discussed.

ACKNOWLEDGEMENT
Financial support from the National Natural Science Foundation of China (NSFC) and the key projects of Beifang University of nationalities (2012XZK02) is gratefully acknowledged.

REFERENCES
[1] G. Magnani, F. Antolini, L. Beaulardi, E. Burresi, A. Coglitore, C. Mingazzini, Sintering, high temperature strength and oxidation resistance of liquid-phase pressureless-sintered SiC–AlN ceramics with addition of rare-earth oxides. J. Euro. Ceram. Soc., 2009, 29(11): 2411-2417.
[2] B. V. Manoj, M. H. Roh, Y. W. Kim, W. Kim, and S. W. Park, Mechanical properties of SiC ceramics sintered with Re_2O_3(Re: Sc, Lu, Y) and AlN additives. Met. Mater. Int., 2010, 16(2): 229-233.
[3] R.M. Balestra, S. Ribeiro, S.P. Taguchi, F.V. Motta, C. Bormio-Nunes, Wetting behaviour of Y_2O_3/AlN additive on SiC ceramics. J. Euro. Ceram. Soc., 2006, 26(16): 3881–3886.
[4] Wu Laner, Chen Yuhong, Jiang Yong, Huang Zhenkun. Liquid sintering of SiC with AlN–Re_2O_3 Additives. J.of the Chinese ceramic society, 2008, 36(5): 593-596.
[5] Z. K. Huang, P. Greil, and G. Petzow, Formation of α-solid solution in the system Si_3N_4-AlN-Y_2O_3 J. Am. Ceram. Soc., 1983, 66(6): 96-97.
[6] Z. K. Huang, T. Y. Tien, and T. S. Yen, Subsolidus Phase Relationships in Si_3N_4-AlN-Rare-Earth Oxide Systems. J. Am. Ceram. Soc., 1986, C69 (10): 241-242.

[7] Z. K. Huang and D. S. Yan, Phase relationships in Si_3N_4-AlN-MxOy systems and their implications for SiAlON fabrication. J. Mater. Sci., 1992, 27: 5640-44.

[8] R. Marchand, Oxynitrides with potassium nickel (II) tetrafluoride structure. Ln_2AlO_3N compounds (Ln = lanthanum, neodymium, samarium). C. R. Acad. Sci. Ser. 1976, C 282 (7): 329-31.

[9] Z. K. Huang and D.S. Yan (T.S. Yen), Compound formation and melting behavior in the AB compound and rare earth oxide systems. J. Solid State Chemistry, 1990, 85: 51-55.

[10] Yuhong Chen, Wenzhou Sun, Laner Wu, Yong Jiang & Zhenkun Huang, Phase Relations in SiC-AlN-R_2O_3 (R = Nd, Gd, Yb, Y) Systems. Journal of Phase Equilibria and Diffusion. 2013, 34(1): 3-8.

[11] Laner Wu, Wenzhou Sun, Yuhong Chen, Youjun Lu, Yong Jiang, and Zhenkun Huang, Phase relations in Si-C-N-O-R (R=La,Gd,Y) systems. J. Am. Ceram. Soc., 2011, 94(12): 4453–4458.

[12] D. P. Thompson, Privative Communication, Newcastle University, Newcastle, UK, 1982.

MICROSTRUCTURE AND PROPERTIES OF REACTION BONDED METAL MODIFIED CERAMICS

S. M. Salamone and M. K. Aghajanian
M Cubed Technologies, 1 Tralee Industrial Park
Newark, DE 19711

S. E. Horner and J. Q. Zheng
Program Executive Office-Soldier, US Army
Ft. Belvoir, VA 22060

ABSTRACT

As a class of materials, reaction bonded ceramics can be readily tailored to obtain the desired properties. Varying the starting powder (preform) characteristics such as morphology, particle size, and composition is one avenue to achieve the required properties. Another method is to alter the infiltration melt. Typical RB(SiC/B$_4$C) composites use liquid silicon as the infiltration metal. However, the infiltration metal can be alloyed with other metallic species to change the composition and properties of the solidified phases. Several Si-(Cu, Fe)/SiC/B$_4$C composites were fabricated using the reaction bonding technique with an alloyed melt infiltration containing as much as 20 wgt% Cu and/or Fe. The infiltrated samples were studied using X-ray diffraction and Scanning Electron Microscopy. Several new phases, including silicides, were formed and the resultant physical properties were measured. Microstructure imaging, along with EDS analysis identified the location of various copper and iron rich regions contained within the composite. As the amount of secondary metal species increases there is a corresponding decrease in the residual free silicon content. The compositional changes also affect the physical properties such as density and Young's modulus as well as the mechanical properties. The following study will correlate the composition and microstructure with the mechanical properties.

INTRODUCTION

Reaction bonded boron carbide materials have many advantages over sintered and hot pressed materials. Aside from the obvious thermal advantages, low temperature processing (average 1500°C versus >2000°C) and minimal displacement characteristics, near net shape processing (i.e. no shrinkage on densification) there is material flexibility associated with infiltrating a porous preform with a liquid metal alloy. Incorporating different metallic species into the infiltration melt can take advantage of the various properties of the additional elements.

Traditional B$_4$C/SiC based reaction bonded materials employ silicon as the infiltration metal, however, many researchers have explored the interface reactions that control the wetting and subsequent infiltration of other metallic species.[1-4] Many final properties can be tailored to fit the specific end use need depending on the metallic element used to alloy with the silicon. Higher temperature applications can be achieved, corrosion resistance and wear resistance can be increased, structural strength and toughness can be improved.[5-7] All these properties take advantage of the new silicide formation and concomitant reduction in residual silicon.

Copper is an interesting material because of the high corrosion resistance and high conductivity properties that it possesses. Copper also fits into the reduced temperature processing of reaction bonded materials because of its low melting point (~1085°C). Since silicon melts at 1414°C, the copper is fluid and will infiltrate with the silicon. Upon cooling, the copper-silicon system also has many intermetallic phases that can be formed. Copper also has a large solubility for boron, which has been shown to be important in improving the wetting behavior of molten copper-silicon alloys in a boron carbide containing composite.[3]

In the present study, iron was also investigated. Because of its affinity for carbon, it is believed that in conjunction with the copper, which has an affinity for boron, the iron will promote reactivity with the carbon species in the B_4C containing sample. The iron can also react with the silicon and form iron silicides. The iron-silicon phase diagram shows many potential phases that can form. Prior X-ray diffraction data showed that the amount of silicon in the final part was reduced to less than 3 weight percent and in some cases it was eliminated. Depending on the processing parameters, a number of iron-silicides as well as iron-aluminum-silicides were formed.

The reactivity with boron and carbon is what led to the recent experiments that eventually combined the copper and iron in the silicon melt. This has the potential of reducing the residual free silicon present in the $Si/SiC/B_4C$ composites. The amount of residual silicon in reaction bonded materials is important because it has been shown to decrease the desired high stiffness and hardness properties of this class of materials.[8-9]

EXPERIMENTAL PROCEDURE

Commercially available SiC and B_4C powder was used as the base material for all the experimental samples investigated. Compacts consisting of the starting powder were combined with specific levels of additional carbon. Preforms were consolidated using these $SiC/B_4C/C$ mixtures, and then infiltrated in a vacuum furnace with molten Si-Cu and Si-Fe to yield $Si-(Cu, Fe)/SiC/B_4C$ ceramic composites.

The phase identification of the initial powder was performed using X-ray diffraction and the average particle size and distribution were measured by a Microtrac S3500 particle size analyzer. The physical properties of the infiltrated composites were measured using several common techniques summarized in Table I. All the microstructures were characterized by examining fracture surfaces using a JEOL JSM-6400 Scanning Electron Microscope. The scanning electron microscope (SEM) images were taken of fracture surfaces in Back-Scattered Mode to differentiate the phases present. Elemental analysis (EDS) was performed in the same SEM after taking the images. The Knoop hardness values were measured on a Shimadzu-2000 hardness tester.

Samples were sent to an independent laboratory to identify compositional changes occurring over the various thermal conditions. The instrument for detecting the different phases present was a Bruker D4 diffractometer with Cu radiation at 45KV/40mA.

Table I: Summary of Properties and Techniques Used to Quantify the Various Composites.

Property	Technique	Standard
Density	Immersion	ASTM B 311
Elastic Modulus	Ultrasonic Pulse Echo	ASTM D 2845
Elemental Analysis	EDS	
Phase Analysis	Powder XRD	
Knoop Hardness	Indentation	

RESULTS AND DISCUSSION

Quantitative phase analysis was conducted on four samples containing copper: the Si/SiC with 10 & 20wt% Cu and Si/SiC/B$_4$C with 10 & 20wt% Cu additions. Table II lists the formulations and the phases detected. The copper in these samples appears to form an ordered intermetallic compound with Al (Cu$_9$Al$_4$), which can be inadvertently introduced in small amounts, during the process, due to the lining material in the alloy container. This type of aluminum interaction has been seen in previous reaction bonding experiments.[10] Only one of the samples (Si/SiC/B$_4$C + 20% Cu) showed a Cu-Si compound (Cu$_7$Si$_2$). This is shown visually in Figure 1 where all of the diffraction patterns are compared over a range showing the peak positions for these two phases. (Note the presence of WC in Table II, this is contamination from the grinding material used at the analytical lab to pulverize the samples for testing.

Table II: Quantitative Phase Analysis of Si/SiC and Si/SiC/B$_4$C Composites with Copper Additions

	Quantitative Phase Analysis (wt%)			
	Si/SiC/B$_4$C +10% Cu	Si/SiC/B$_4$C +20% Cu	Si/SiC +10% Cu	Si/SiC +20% Cu
SiC(6H)	57.6(5)	62.1(6)	85.5(5)	84.9(5)
SiC(15R)	3.5(1)	4.2(1)	4.0(2)	3.9(1)
B$_{13}$C$_2$	9.2(2)	10.1(3)	0.4(1)	
Si	4.6(1)	3.6(1)	6.5(2)	5.0(2)
C			0.3(1)	0.9(2)
WC	1.4(2)	1.6(2)	1.2(1)	1.3(2)
Cu	<0.1	<0.1	<0.1	<0.1
Cu$_9$Al$_4$	2.1(2)	4.2(1)	1.3(2)	3.3(2)
Al$_{0.84}$B$_{39.8}$C$_4$	14.7(5)	12.6(1)	0.7(2)	
AlB$_{24}$C$_4$	6.1(2)			
Al$_4$Si$_4$C$_7$	0.7(1)	0.9(1)		0.6(1)
Cu$_7$Si$_2$		0.6(1)		

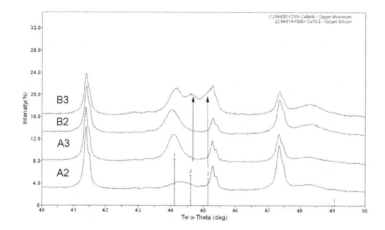

Figure 1: Overlapping diffraction pattern showing the formation (and absence of) the Cu$_7$Si$_2$ (#2) and the Cu$_9$Al$_4$ (#1) intermetallic phases.

After the initial copper experiments, additions of copper, iron and a mixture of copper and iron were added to the silicon metal and preforms of SiC/B$_4$C/C were infiltrated. Table III lists the physical properties of the various composites. As with the copper only additions, as the amount of copper and iron is increased the density increases, but the Young's modulus decreases or is constant. The physical properties of the 10% iron composite are slightly higher in both density and modulus than the 10% copper composite. The densities for copper and iron are similar (Cu-8.96 g/cc & Fe-7.87 g/cc), however, the Young's modulus of iron is roughly twice that of copper. All the samples were fully dense, as the microstructures will show. Samples with higher weight percentages were fabricated but not all the various compositions were fully dense. Some showed signs of porosity so they were not included in the property data or the analysis.

Table III: Property Data for Si/SiC and Si/SiC/B$_4$C with Copper and Iron Additions

Composition (wt %)	ID	Density (g/cc)	Young's Modulus (GPa)
Si/SiC	A1	3.09	404
Si/SiC +10% Cu	A2	3.13	408
Si/SiC +20% Cu	A3	3.18	410
Si/SiC/B$_4$C	B1	2.91	422
Si/SiC/B$_4$C +10% Cu	B2	2.94	420
Si/SiC/B$_4$C +20% Cu	B3	2.99	418
Si/SiC/B$_4$C +5% Cu & 5% Fe	B4	2.96	426
Si/SiC/B$_4$C +10% Cu & 10% Fe	B5	3.04	425
Si/SiC/B$_4$C +10% Fe	B6	2.96	431

The microstructures of representative samples are shown in Figure 2. Back scattered images highlight the contrast between different phases in a composite system. The dark regions are the grains of B$_4$C and the grey areas are SiC grains. The light regions are silicon metal and the extremely bright areas in the images are the copper and iron rich regions. The metallic phases appear to be dispersed throughout the samples with a few localized regions of high concentration. However, these regions are smaller than the grain size of the Si-(Cu, Fe)/SiC/B$_4$C matrix particles.

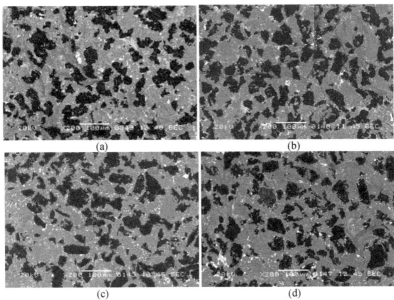

(a)   (b)

(c)   (d)

Figure 2: Back-scattered SEM images of the fracture surfaces of Si/SiC/B$_4$C with (a) 10% Cu, (b) 5% Cu & 5% Fe, (c) 10% Cu & 10% Fe and (d) 10% Fe additions

Elemental analysis was performed on the samples to confirm the presence of copper and iron in the composites. Although there is no quantitative phase analysis (e.g. crystal structure of silicide formation) the atomic percent of each element detected is given. Figure 3 is a compilation of EDS spectra showing the presence of copper and/or iron in the bright regions shown in the inset microstructures highlighted with a red circle. The other regions were probed and the presence of boron, silicon and carbon were found in the expected atomic ratios.

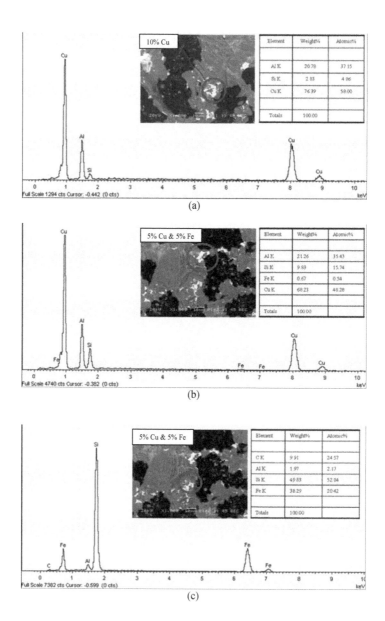

(a)

(b)

(c)

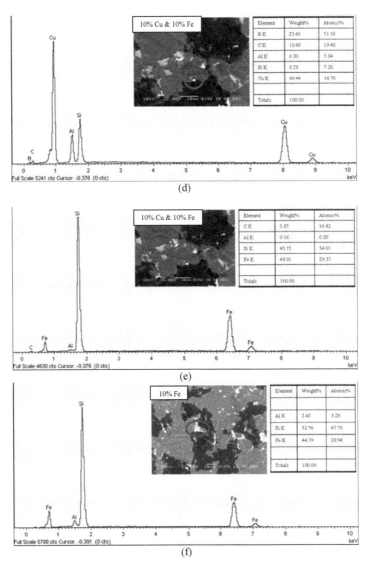

Figure 3: EDS spectra of the Si/SiC/B₄C samples containing (a) 10% Cu, (b & c) 5% Cu & 5% Fe, (d & e) 10% Cu & 10% Fe and (f) 10% Fe additions.

The hardness values of the Si/SiC/B$_4$C composites with varying amounts of copper and/or iron are shown in Figure 4. In all cases, the addition of a copper and/or iron species to the silicon melt increased the hardness as compared to the baseline Si/SiC/B$_4$C composite. Several interesting observations can be made based on the resultant hardness values. As the amount of copper is increased, the hardness value increases. The addition of 10% copper has virtually the same effect as the addition of 10% iron on the hardness value, i.e., an increase of about 7.5 percent. Mixing of the two metallic species has an even greater effect than the addition of the single element. By adding 5%Cu and 5% Fe the hardness is higher than adding either the copper or iron only. The same is true for adding 20 percent (10% of each element). The XRD data suggests that as the metallic additions increase, the silicon content decreases and it has been shown that the hardness increases with decreasing silicon content in reaction bonded boron carbide materials.[11-12] (Each bar is an average of five hardness measurements.)

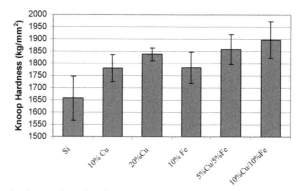

Figure 4: <u>Knoop hardness values showing an increase in hardness with the addition of Cu and/or Fe.</u>

SUMMARY

The fabrication of fully infiltrated Si-(Cu, Fe)/SiC/B$_4$C composites with up to 20 weight percent of Cu and/or Fe added to the silicon melt has been demonstrated. The Cu/Fe addition increases the density of the Si/SiC/B$_4$C composites, but does not significantly affect the Young's modulus. However, the hardness, as measured by a Knoop indenter, is increased for all combinations and amounts of copper and iron added. The combination of copper and iron also has a greater influence on the hardness than just the addition of copper or iron alone.

ACKNOWLEDGEMENT

This work was funded by Program Executive Office-Soldier, US Army, under contract number W91CRB-11-C-0085.

REFERENCES
[1]N. Froumin, N. Frage, M. Aizenstein, and M. Dariel, "Ceramic-metal interaction and wetting phenomena in the B$_4$C/Cu system", *J. Euro. Ceram. Soc.*, **23**, 2821-2828, (2003).

[2]M. Aizenstein, N. Froumin, N. Frage, and M. Dariel, "Interface phenomena in the B$_4$C/(Me-Ti) systems (Me=Cu, Au and Sn)", *J. Mater. Sci.*, **40**, 2325-2327, (2005).

[3]N. Frage, N. Froumin, M. Aizenstein, and M. Dariel, "Interface reaction in the B_4C/(Cu-Si) system", *Acta Materialia*, **52**, 2625-2635, (2004).

[4]M. Singh and D.R. Behrendt, "Reactive melt infiltration of silicon-niobium alloys in microporous carbons", *J. Mater. Res.*, **9,** [7], (1994).

[5]I. Mizrahi, A. Raviv, H. Dilman, M. Aizenstein, M. P. Dariel and N. Frage, "The effect of Fe addition on processing and mechanical properties of reaction infiltrated boron carbide-based composites," *J. Mater. Sci.*, **42**, [16], 6923-6928, (2007).

[6]R. Messner and Y-M. Chiang, "Liquid-Phase Reaction-Bonding of Silicon Carbide Using Alloyed Silicon-Molybdenum Melts," *J. Am. Ceram. Soc.*, **73**, [5], 1193-1200, (1990).

[7]S. Tariolle, F. Thevenot, M. Aizenstein, M. Dariel, N. Froumin, and N. Frage, "Boron carbide-copper infiltrated cermets", *J. Sol. St. Chem.*, **177**, 400-406, (2004).

[8]S. Salamone, P. Karandikar, A. Marshall, D. D. Marchant, M. Sennett, "Effects of Si:SiC Ratio and SiC Grain Size on Properties of RBSC", in Mechanical Properties and Performance of Engineering Ceramics and Composites III, *Ceram. Eng. Sci. Proc.*, **28**, E. Lara-Curzio et al. editors, 101-109, (2008).

[9]O. P. Chakrabarti, S. Ghosh, and J. Mukerji, "Influence of Grain Size, Free Silicon Content and Temperature on the Strength and Toughness of Reaction-Bonded Silicon Carbide", *Ceramics International*, **20**, 283-286, (1994).

[10]S. Hayun, H. Dilman, M. P. Dariel and N. Frage, "The effect of aluminum on the microstructure and phase composition of boron carbide infiltrated with silicon", *Materials Chemistry and Physics*, **118** [2-3], 490-495, (2009).

[11]M. Dariel and N. Frage, "Reaction bonded boron carbide: recent developments", *Adv. Appl. Ceram.*, **111**, [5-6], 301-310, (2012).

[12]P. Chhillar, M. K. Aghajanian, D. D. Marchant, R. A. Haber, and M. Sennett, "The effect of Si content on the properties of B_4C-SiC-Si composites," CESP, **28**, [5], 161-167, (2007).

INVESTIGATION INTO THE EFFECT OF COMMON CERAMIC CORE ADDITIVES ON THE CRYSTALLISATION AND SINTERING OF AMORPHOUS SILICA

Ben Taylor, Stewart T Welch and Stuart Blackburn
School of Metallurgy and Materials, Birmingham University, Edgbaston, Birmingham, UK
Precision Casting Facility, Rolls-Royce PLC, Derby, UK

ABSTRACT

Aero engine turbine blades are commonly produced via investment casting methods and utilise sacrificial ceramic cores to provide internal features such as cooling channels. During the firing process the conversion of the main ingredient (amorphous silica) into β-cristobalite plays a significant role, as it directly affects the dimensional stability, shrinkage and leachability of the core after casting. The formulation used to produce ceramic cores has evolved over the years in an iterative fashion, resulting in a deficit of understanding regarding the role of each component in the now complex formulation. Dilatometry was utilised to evaluate common additives to the amorphous silica system, such as zirconium silicate (zircon), alumina, aluminosilicate and magnesium oxide. Green bodies were prepared via die compaction, then pre-fired to 1200°C combining binder burnout with an initial sintering phase to give the ceramic mechanical strength. Subsequently samples were tested up to temperatures typical for casting (above 1500°C). Comparisons were made of the dimensional properties of the ceramic formulations during the pre-fire and the casting cycle. It was shown that the addition of aluminosilicate and zirconium silicate provides the desired dimensional stability during a simulated casting cycle.

INTRODUCTION

Investment casting provides a route for the manufacture of highly complex gas turbine components. These components are becoming increasingly elaborate in design requiring intricate cooling channels and finer feature sizes. In the investment casting process a disposable ceramic core most often manufactured by injection moulding is used to form these cooling channels (Figure 1). The core is placed in a die and wax injected around it to produce a wax pattern replicating the shape of the turbine component. This 'wax pattern' can then be added with other similar component wax patterns to a runner to make a tree or pattern assembly. This system is coated with a 'shell' by first dipping in fine ceramic slurry (prime coat), which is then stuccoed with a coarse grained particulate to build volume. The process is repeated until the desired thickness is achieved, with the binder system granting mechanical strength. To dewax the assembly autoclaving or rapid firing compensate for the much greater thermal expansion coefficient of the wax compared to the shell by rapidly heating the system. This initial melting causes the outer layer of the wax to melt and absorb into the shell or drain to allow room for the bulk of the wax to expand and melt, thus avoiding cracking and failures in the mould. Minor flaws can be repaired at this stage before a pre-sintering or pre-firing process is undertaken to burnout residual wax / binder and to harden the mould leaving it ready for casting (Figure 2).

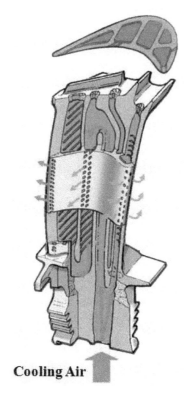

Figure 1 – Cross sectional schematic of a multi-pass ceramic core within a turbine blade, providing airflow channels for cooling[1]

Nickel based superalloys are introduced to the mould at temperatures above 1500°C, and then cooled in a highly controlled environment generating a single-crystal, directionally solidified or equiaxed components depending on mould and furnace features. The ceramic shell is removed and the core leached in alkali solutions before the final finishing and inspection processes.

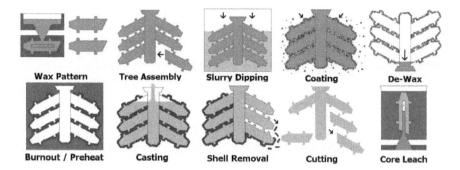

Figure 2 – Schematic of the Investment Casting Process

It is advantageous to have a ceramic core formulation that is engineered to withstand the casting process at the highest possible temperature. This allows alloys with high melting points to be used, increasing the operating temperature of the engine and thus improving efficiency and thrust. Single-crystal castings provide the best solution, however the complexity of the process yields many constraints on the formulation of the ceramic core. Some of the requirements were stated in a paper by Wereszczak et al. and are as follows:[2]

- "it must be chemically compatible with the metal;
- be much weaker than the metal so that it will crush (i.e., exhibit good "crushability") during metal solidification rather than impose deleterious and destructive high stresses (e.g., those that lead to secondary crystallization in single crystal castings, grain boundary cracking in directionally solidified castings, or hot-tearing in equiaxed castings) into the metal;
- have enough strength to be resistant to thermal shock;
- be sufficiently resistant to dimensional changes (i.e., exhibit good "stability") and yield a suitable metal surface finish; and
- be removable by a process that is not harmful to the metal casting".

The possible choice of materials for a ceramic core is limited by the need to meet the above requirements. A ceramic core formulation tends to be based on vitreous silica, alumina and zircon, potentially with other oxides added to improve performance. Silica is used as it is one of the few materials which can be removed with moderate ease via leaching without affecting the casting detrimentally. The green ceramic core along with the rest of the mould undergoes a pre-fire phase to intermediate temperatures (circa 1200°C) before being used in the high temperature cast cycle (above 1500°C). During the various heating processes applied to the ceramic the component materials utilised undergo phase transformations, especially in the casting temperature range, and therefore change density and mechanical properties while doing so. The materials produced in this process include cristobalite[3], a polymorph of silica, and mullite[4], an aluminosilicate. The formation rate of these materials and the temperatures at which these transitions occur is controlled by many factors, including ceramic formulation, hydroxyl content, and the manufacturing conditions and method employed. The mechanical properties of the final ceramic product are also influenced dramatically by slight changes in the above parameters[5], and understanding these factors better is of particular interest as a model that can predict the fundamentals and kinetics of phase development and sintering does not exist, and currently a trial and error approach is employed to obtain the desired ceramic properties.

It is imperative that during the pre-fire stage an amount of cristobalite is formed ($\approx$10-30%).[6] This initial level of cristobalite forms a skeletal structure within the ceramic body to provide not only initial strength to allow handling, but also a foundation for further cristobalite growth during the high temperature casting cycle to provide resistance to dimensional change.[7] The cristobalite produced acts to counteract the shrinkage during sintering by the volumetric expansion during devitrification, and by pinning the movement of grain boundaries as it forms on the surface of grains by nucleation and growth. If too much cristobalite is formed during the pre-fire however, substantial cracking on cooling, prior to being reheated to cast temperature, can cause the ceramic to fail. There is a significant volume change (3.2%) that takes place during the β-α cristobalite phase change. This phase change occurs around 250°C and is dependent on the size of the grains in the sintered product. Smaller grain sizes possess greater stability and can persist to lower temperatures, or when the structure is 'stuffed' with large cations the phase change may not occur at all.[8] This phase change becomes beneficial after casting as the volume change induces cracking which provides a network of porosity which improves the ease and speed of leeching.

The reconstructive phase change from silica glass to β-cristobalite (above 1000°C typically) occurs over a period of time, and involves the gradual reordering of SiO_4 tetrahedra. During this phase transformation there is a reduction in density with associated volumetric expansion due to adopting the more open structure of β-cristobalite. One way for this to occur is by nucleation and growth in the solid state. Another process acts when there is a an appreciable vapour pressure, amorphous silica may vaporise and condense as the more stable, lower vapour pressure β-cristobalite form.[9] Transformations are speeded up by the presence of a liquid phase providing solubility to allow amorphous silica to become a solution and then precipitate as β-cristobalite. This process is utilised in most ceramic industries, mineralisers and fluxes such as group I alkali metals are added to disrupt the structure and form a liquid phase at lower temperatures. The α-β cristobalite phase change is displacive and requires only a change in the secondary coordination of the atoms and requires no bonds be broken. As such the transition has a particular activation energy and occurs very rapidly at a given temperature.[9] The high temperature cubic β-cristobalite phase has the highest symmetry, is the most stable and once formed can persist metastably at temperatures above $\approx$250°C with a density of 2.26 g/cm^3. The low temperature tetragonal modification α-cristobalite is structurally distorted derivative with a higher density of 2.33 g/cm^3 as shown in Figure 3.

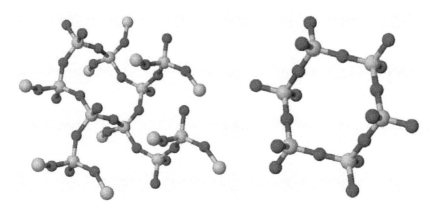

Figure 3 – Idealised structure from left to right; α-cristobalite and β-cristobalite[10]

EXPERIMENTAL TECHNIQUES

Materials Used

All powders used are single phase high purity products for applications in advanced ceramics. Materials include amorphous fused silica, zirconium silicate, MgO, alumina, and aluminosilicate.

Particle Size Analysis

Particle size analysis was undertaken using a Sympatec 'Helos' particle size and shape analyser using three different optical lenses to study different particle size ranges, and a 'Rodos' dry air dispersion system to avoid agglomeration of particles which would affect the results. Powders were loosely ground with a pestle and mortar to break up any large agglomerates prior to analysis. The total measuring range utilising all three optical systems ranges from 0.18 to 875 μm (Lens 1 = 0.18-35 μm, Lens 2 = 0.9-75 μm, Lens 3 = 4.5-875 μm). Typically Lens 1 is utilised when evaluating fine grade material.

Milling

Powders were milled using an attrition mill with zirconia stirrer system and media. The container was made from a polymer which would burn out at low temperatures if low level contamination did occur. The time in the attrition mill depended on the toughness and initial grade of the material, with the purpose of producing different materials with the same particle size distribution. These powders are all given the code SF in the following results and discussion.

Powder Formulation Blending

A 'DAC 150' speedmixer was employed for mixing powders blends which utilises a dual asymmetric centrifuge to thoroughly mix materials. Mixing was undertaken in isopropanol with the addition of 5, 5 mm zirconia beads to encourage dispersion of the powder. The material was mixed for 15 minutes, then dried in an oven and ground with a pestle and mortar to de-agglomerate the material.

Green Body Formation

Cuboid tablets measuring 5.5 mm x 40 mm x 20 mm were produced by pressing the various powder formulations using a bespoke die press to 7.5 MPa, with a binder consisting of PEG 200 and water (3:1 ratio) to aid lubrication.These tablets were then fired to 1200°C in a cycle that incorporates both debinding and pre-firing the ceramic to allow handling and cutting. The mass and basic geometric measurements were recorded before and after the pre-fire. Dilatometer samples measuring 5.5 mm x 5.5 mm x 20 mm were cut from the formed tablets using a Dremel® saw, with samples under scrutiny only selected from the centre section of the tablet to minimise density variation caused during sample preparation.

Dilatometry

Dilatometry was performed using a Netzsch DIL 402PC horizontal pushrod dilatometer, allowing measurement of expansion and contraction over a programmable temperature / time profile. Correction runs where undertaken prior to analysis using a densified alumina test sample. The mass and basic geometric measurements were recorded before and after testing. The 'cast cycle' was carried out by the dilatometer, and involved first a 10°C min^{-1} ramp up to 750°C with a 20 minute hold (this was to get the sample up to temperature and equilibrate) followed by a selected ramp up to 1530°C (typically between $4 - 12$°C min^{-1}). The sample was then cooled back to room temperature at a controlled rate of 10°C min^{-1}. Variation from this cooling rate may have occurred at lower temperatures as the system does not provide forced cooling. Data is shown for both the ramp up and the ramp down as a single curve in the presented results. Also the number of data points is significantly higher than the number of points on the curve; this has been reduced for ease of viewing.

RESULTS AND DISCUSSION

Some provisional work was undertaken to optimise the selected test regimes using pure silica of different sizes. This provided a baseline of knowledge before moving to a more complex system. A summary of the effect of key variables in the study is provided:

- Compaction force applied when pressing samples is insignificant
- Particles size distribution is significant, also with screening of fine particulates (<10μm) higher levels of shrinkage are obtained due to lower initial density
- Slower heating cycle (longer in furnace) reduces shrinkage and more cristobalite is produced
- Cristobalite formation limits shrinkage due to intrinsic expansion and formation of skeletal structure within ceramic body
- Sintering requires less thermal energy than crystallisation

The preliminary investigation concerned a variety of materials which are typically used in ceramics and the investment casting industry to influence the behaviour of the pure silica system. Powders were milled to near equal size to make the data more representative of the chemistry of the material in question (Table 1). The milled powders are labelled superfine (SF) in the results.

These powder blends where then tested via dilatometry at various heating rates. For simplicity only the 12°C min⁻¹ ramp rate data are shown as a comparison.

Material	Before Milling (AR)			After Milling (SF)			Comparison		
	D10	D50	D90	D10	D50	D90	D10	D50	D90
Silica	1.23	19.02	95.71	-	-	-	1.23	19.02	95.71
Alumina	1.19	3.25	6.56	0.45	1.59	3.22	0.45	1.59	3.22
Zircon	1.11	13.65	40.49	0.31	1.37	3.22	0.31	1.37	3.22
MgO	0.40	1.54	3.52	-	-	-	0.40	1.54	3.52
Aluminosilicate	0.47	2.16	8.67	-	-	-	0.47	2.16	8.67

Table 1 – Particle size distribution of as received (AR) and milled (SF) powders

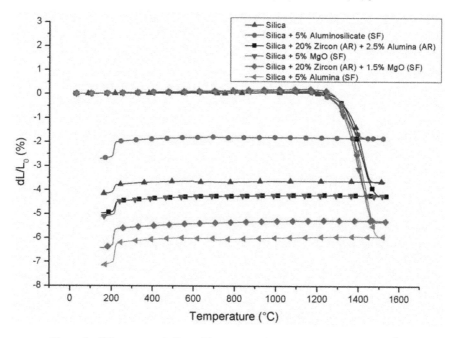

Figure 4 – Dilatometry of silica with common additives (ramp rate 12°C min⁻¹)

The data in Figure 4 although showing limited formulations illustrates that to obtain less shrinkage than the baseline silica system, aluminosilicate needs to be added. Additions of alumina and magnesia make the system densify to a higher degree; the formation of cristobalite is not able to compensate for the oxide dopant additions in terms of shrinkage inhibition. A 5% addition of alumina induces higher shrinkage than a 5% addition of magnesia, which is in agreement with the liquidus temperature for the corresponding phase diagrams.[9] The alumina based system has a lower liquidus temperature therefore the system will have lower viscosity at high temperatures increasing the sintering rate. However in the presence of zircon, magnesia doping induces higher shrinkage that alumina doping, a reverse trend. The effect of combining alumina and MgO with zircon has the opposite effect regarding the shrinkage for each material, but still act to increase the rate of sintering beyond the rate of crystallisation at elevated temperatures. This research led to further investigation of the silica, zircon, aluminosilicate system, to further promote a reduction in observed shrinkage. The dimensional step change occurring around 200°C during cooling is related to the cristobalite produced during the final heating cycle.

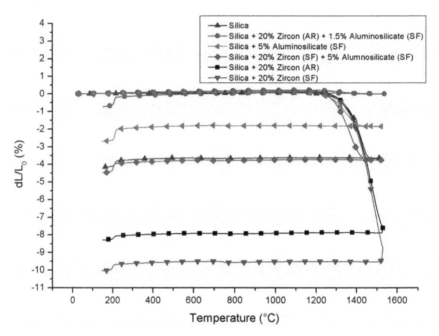

Figure 5 – Silica, zircon, aluminosilicate system with control on the zircon particle size distribution (ramp rate 12°C min^{-1})

To investigate further the interactions of zircon, aluminosilicate and silica a series of experiments were undertaken using different grades and concentrations, the dilatometer results are presented in figure 5. Selecting the different particle size distributions of the starting powders helped to deconvolute the effect of particle size distribution, from the effect of the chemistry. By adding coarse zircon and low volumes of aluminosilicate almost no shrinkage occurred. With fine zircon and more aluminosilicate greater shrinkage occurred making it equivalent to the pure silica system. Without the presence of aluminosilicate zircon causes higher shrinkage than pure silica, and fine zircon increases this shrinkage further. This is associated with fine material aiding the rate of sintering due to higher surface energy. However the findings that relate to the presence and size of zircon were unexpected. The data suggests that zircon does act to pin grain boundary movement as suggested previously, but only in the presence of aluminosilicate. Also contrary to current understanding[11] there is an inverse correlation to the Zener effect, which defines that smaller particle size acts to increase the drag force of the grain boundary passing across the particle. The coarse zircon may be acting in a similar way to cristobalite, offering an additional internal skeletal structure with low reactivity, however this is not completely understood and scanning electron microscopy studies will be undertaken to observe the development of microstructure.

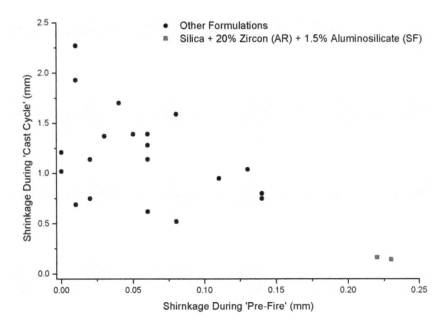

Figure 6 – Scatter plot of shrinking during 'pre-fire' and 'cast cycle' for all formulations at both 4°C min^{-1} and 12°C min^{-1} ramp rates

Figure 6 represents the change in measurements taken along the 20 mm length axis of the tablets before and after the pre-fire, which was repeated over the same axis of the corresponding dilatometer rod during the cast cycle. Data is shown for both the 4°C min^{-1} and 12°C min^{-1} heat cycles. The scatter plot illustrates a trend in which samples with increased shrinkage during the 'pre-fire' step tend to have improved dimensional stability during high temperature sintering equating to a typical cast temperature. This is emphasised by the silica + 20% zircon (AR) + 1.5% aluminosilicate (SF) formulation labelled. This trend agrees with the provisional work on the pure silica system. If sintering and crystallisation reactions take place at a lower temperature due to the chemistry or morphology of the materials present, a structural basis containing cristobalite is formed before the casting cycle. This acts to provide a seed for further cristobalite production and the volumetric expansion during devitrification counteracts the shrinkage during sintering. It is also noteworthy that the upper right area of the graph remains empty. This suggests that for the thermal cycles applied in this work it is not possible to have high shrinkage for both the pre-fire and cast cycle, and signifies the importance of the thermal history of a ceramic and requires further investigation. Figure 7 looks closer at the dilatometer curve and reveals that alumina silicate is fluxing the cristobalite during the pre-fire for this formulation, and can be detected by the small step in expansion caused by the α-β phase transformation during heating at ≈250°C. This may be seeding the further growth of cristobalite through the high temperature firing which is currently under investigation by x-ray diffraction. Greater densification is avoided by the presence of the course zircon while the finer zircon aids sintering acting to counter the cristobalite formation and increase shrinkage.

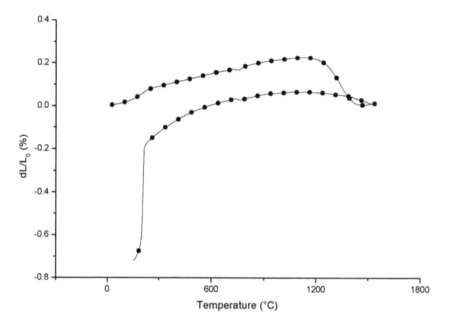

Figure 7 – Dilatometry plot of cast cycle for silica + 20% zircon 200 + 1.5% aluminosilicate SF formulation

CONCLUSIONS

A comparison of a range of formulations based on amorphous silica was conducted by means of dilatometry. The study revealed that using coarse zircon and additional fine aluminosilicate as additives provides minimal shrinkage during the high temperature cast cycle, which is desirable for applications in investment casting. This is shown to be due to production of cristobalite at low temperatures, supporting the notion that cristobalite formation helps provide an internal structure to the ceramic that resists dimensional change. A trend is observed for all formulations that helps confirm this theory, illustrating that higher shrinking during the pre-fire tends to low shrinkage during the cast cycle.

ACKNOWLEDGEMENTS

The authors would like to thank the University of Birmingham for the provision of the facilities used in the preparation of this paper. Further they thank the EPSRC and Rolls-Royce Plc for their financial support.

REFERENCES

1. *Rolls Royce Plc.*
2. Wereszczak, A.A., et al., *Dimensional changes and creep of silica core ceramics used in investment casting of superalloys.* Journal of Materials Science, 2002. **37**(19): p. 4235-4245.
3. Huseby, I.C., M.P. Borom, and C.D. Greskovich, *High-Temperature Characterization of Silica-Base Cores for Super-Alloys.* American Ceramic Society Bulletin, 1979. **58**(4): p. 448-452.
4. Saruhan, B., et al., *Reaction and sintering mechanisms of mullite in the systems cristobalite/alpha-Al2O3 and amorphous SiO2/alpha-Al2O3.* Journal of the European Ceramic Society, 1996. **16**(10): p. 1075-1081.
5. Chao, C.H. and H.Y. Lu, *Optimal composition of zircon-fused silica ceramic cores for casting superalloys.* Journal of the American Ceramic Society, 2002. **85**(4): p. 773-779.
6. Miller, J.J., *Cores for Investment Casting Process.* Sherwood Refractories, Inc., 1978(United States Patent, 4,093,017).
7. Wang, L.Y. and M.H. Hon, *The Effect of Cristobalite Seed on the Crystallization of Fused-Silica Based Ceramic Core - a Kinetic-Study.* Ceramics International, 1995. **21**(3): p. 187-193.
8. Lee, S.J. and C.H. Lee, *Critical size effect for chemically doped b-cristobalite Transformation.* Materials Letters, 2000. **45**: p. 175-179.
9. Kingery, W.D., H.K. Bowen, and D.R. Uhlmann, *Introduction to ceramics.* 2nd ed. / [by] W.D. Kingery, H.K. Bowen, D.R. Uhlmann. ed. 1976, New York ; London: Wiley-Interscience. xiii,1032p.
10. *From Jmol: an open-source Java viewer for chemical structures in 3D. http://www.jmol.org/.*
11. Wilson, P.J., et al., *The role of zircon particle size distribution, surface area and contamination on the properties of silica-zircon ceramic materials.* Journal of the European Ceramic Society, 2011. **31**(9): p. 1849-1855.

DIFFERENT FIBERS EXPOSED TO TEMPERATURES UP TO 1000° C

Henry A. Colorado[1,2], Clem Hiel[3], Jenn-Ming Yang[1]

[1]Materials Science and Engineering Department, University of California, Los Angeles, CA 90095, USA
[2]Universidad de Antioquia, Mechanical Engineering. Medellin-Colombia
[3]Composite Support and Solutions Inc. San Pedro, California. Mechanics of Materials and Constructions. University of Brussels (VUB)

ABSTRACT

E-glass, Carbon, Basalt and SiC fibers were exposed to high temperature in air environment. Samples were placed in a furnace at different temperatures (up to 1000°C) to determine the thermal effect in the tensile strength using the single-fiber testing method. Weibull statistics were made for all fibers and exposure temperatures. The characterization of the microstructure was conducted by scanning electron microscopy, x-ray diffraction and thermo-gravimetric analysis. Results showed important aspects that may be considered when these fibers are reinforcing ceramic composites under high temperature oxidation environments. It was found that for all fibers the tensile strength was significantly affected by the exposure temperature.

INTRODUCTION

The increase in the requirements and applications for the thermal and chemical stability of composites under high temperature and oxidation environments has stimulated the development of many ceramic and polymer matrix composites with fibers. However, under high temperature and oxidation exposure even ceramic fibers show significant decrease in their properties.

It has been established that the most common reinforcement used in polymer matrix composites is fiberglass. Approximately 90–95% of all composite products contain glass fibers[1]. In previous studies for glass fiber degradation[1] at temperatures up to 650°C it has been found that the strength degradation is a result of larger surface flaws present after heat treatment. Tests were mostly done using the fiber bundle test and only one temperature effect (450°C) was investigated with the single-fiber test. The tensile strength of glass fibers decreases as temperature increases and this has been attributed to several mechanisms such as (1) the annealing of compressive residual stresses, (2) the re-orientation or the loss of orientation of a microstructure, (3) the formation of a surface layer with different properties to the fiber core, and (4) the development of surface flaws due to phenomena such as the corrosion at high-temperature. Mechanisms (1) and (3) are well accepted. Mechanisms (2) and (4) are the most popular, despite little evidence of either network orientation or differences between the surface and core properties of the fiber. Thus, more research has to be conducted in order to explain this complex process. Moreover, when glass fiber is in a ceramic environment in addition to be exposed to oxidation, depending on the temperature, it is in an active inter-diffusion environment, which certainly makes the analysis more complex.

Carbon fiber oxidation has been studied before under diverse thermal conditions[2] and it has been found that the strength decreases with the temperature and the surface oxidation is very uniform. Therefore, the mechanical strength of the fibers also falls as the fiber size decreases. Equations 1 and 2 show the chemical reactions that lead the oxidation. It is well known in composites that once the interfacial bond between the fiber surface and the matrix has been debonded by oxidation, a significant reduction in the mechanical properties of the composite is obtained. Therefore, it is very important to understand the mechanisms of that deterioration of singles fibers in order to not only explain the composite behavior but also to decrease such oxidation.

$$C + O_2 = CO_2 \qquad\qquad (1)$$
$$C + 0.5\ O_2 = CO \qquad\qquad (2)$$

Basalt fiber oxidation at different temperatures and its problem of crystallization (even during the manufacturing) of some of the components (which limits it applications to temperatures below 800°C) has been studied before[3]. However, due to the relatively new use of these fibers in non-military applications (after 1995), there is very few data and as a consequence for the full understanding of the mechanisms involved in the deterioration due to the oxidation of the fibers at high temperatures needs to much research. Due to the abundance of the raw materials (basalt rock is the most wide spread volcanic stone in the earth's crust), the have obtained a lot of attention to be used in diverse composites.

Silicon carbide oxidation at different temperatures has been well documented before[4]. Many byproducts can be obtained depending on the chemical reaction. Some of the possible reactions are summarized in equations 3-6.

$$SiC + 4O \rightarrow SiO_2 + CO_2 \qquad\qquad (3)$$
$$SiC + 2O \rightarrow SiO + CO \qquad\qquad (4)$$
$$SiO_2 + C \qquad\qquad (5)$$
$$SiC + O \rightarrow Si + CO \qquad\qquad (6)$$

In this paper, some of the most popular fibers for industry (E-glass, carbon, basalt and SiC) have been studied under oxidation environments at different temperatures (200°C, 400°C, 600°C, 800°C, and 1000°C) and their properties have been compared before and after the thermal processes. The paper shows comparative results and it focus in the mechanical behaviour. This is because since many studies have been conducted with the bundle fiber method for these fibers (which is easier to conduct because of the sample preparation) there is few data about regarding Weibull parameters for single fibers, which are very different when they are compare to the bundle fiber results. Further motivation for this research is to develop in future inexpensive methods to protect these fibers in extreme oxidation environments such as extreme pH environments, oxidation due to high temperature exposure and thermal shock.

EXPERIMENTAL PROCEDURE

Glass fibers from Textrand 225 from Fiber Glass Industries, Graphite fibers Tenax(R)-A 511, Basalt fibers BCF13-1200KV12 Int from Kammemy Vek, and SiC fibers Nicalon[TM] were used in this research.

All samples were annealed for 1h at 200°C, 400°C, 600°C, 800°C, and 1000°C. In order to observe the cross section of the fibers at SEM (JEOL JSM 6700R), fibers were put into resin and then polished (after sputtered in a Hummer 6.2 system at conditions of 15mA AC for 30 sec to obtain a thin film of Au) until mirror finishing using SiC papers and alumina of 1μm. At least 20 fibers at each thermal condition were measured in diameter in order to build up size distributions. Not all fibers types were able to be characterized at all thermal treatments due to diverse problems such as fiber stability and handling. Over some temperatures some fibers either disappeared completely or deteriorated in such a way that it was not able to prepare a sample for tensile tests.

Fibers were ground for X-Ray Diffraction (XRD) experiments, which were conducted in an X'Pert PRO (Cu Kα radiation, λ=1.5406 Å) equipment, at 45KV and scanning between 10° and 70°. For each type of fiber and thermal treatment, twenty samples were tested by tensile tests in order to plot Weibull distributions. These tests were performed on an Instron 4411 at a crosshead speed of 2.5 mm/min. Thermo gravimetric Analysis (TGA) was performed in Perkin Elmer Instruments Pyris Diamond TG/DTA equipment. The temperature ramp was 10°C/min from room temperature up to 1000°C, with previous equilibration at 50°C for 10min. All experiments were conducted in air atmosphere in an alumina pan with air at 100mL/min.

RESULTS

Figure 1shows SEM images of the fibers' surface at some different temperatures. Figure 1a some shows glass fibers as received and after 600°C oxidation. After the 600°C oxidation treatment the fibers weakens a lot that even regular handling to use them at SEM has produced the cut shown in the image. Figure 1b shows carbon fibers as received and after 600°C oxidation. It is observed how the fiber diameter reduction and both surface and core damage are significant. Figure 1c shows basalt fibers as received, after 600°C and after 1000°C oxidation. It is observed the progressive formation of an oxide layer as temperature of the oxidation is increased. Figure 1d shows SiC fibers as received, after 600°C and after 1000°C oxidation. It is observed that after 600°C the fiber looks without damages and in fact it seems even better that the as received fiber, which can be due to a possible cleaning of the fiber surface. At 1000°C fibers in general show small deterioration, however some of them as shown in the image have internal and surface damages.

a) Glass fibers as received   Glass fibers exposed to 600°C   Glass fibers exposed to 600°C

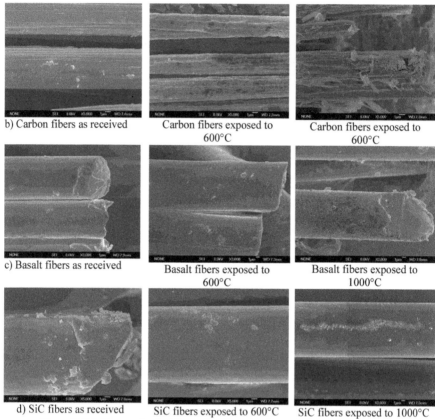

Figure 1 SEM images for the fibers after the oxidation at different temperatures.

Figure 2 shows the cross section view for the fibers after 400°C oxidation. For glass, carbon, basalt and SiC fibers the shown magnification was always used for the same fiber at all temperatures in order to measure the diameter in the SEM.

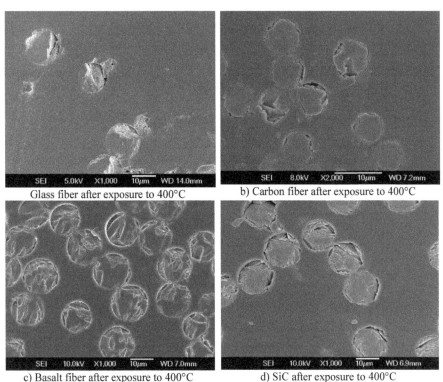

Glass fiber after exposure to 400°C

b) Carbon fiber after exposure to 400°C

c) Basalt fiber after exposure to 400°C

d) SiC after exposure to 400°C

Figure 2 SEM cross section view images for the fibers after exposure to 400°C.

Figure 3 shows different diameter distributions for the fibers. In general it is observed that as the temperature of the thermal oxidation treatment increase, the variability of the size decreases as well, which is due to a very different phenomena as we will see below with the other characterization. For glass and basalt fibers at 800 and 1000°C respectively this variability has decreased. Carbon fibers also decreased this variability after the 400°C oxidation. SiC fibers also decreased this variability in diameter. It has been also observed that for almost all fibers as temeprature increased, fiber diameter increased in most of the cases, which is due to structural factors (such as void formation, melting or internal deterioation) or due to the increase of oxides in the surface.

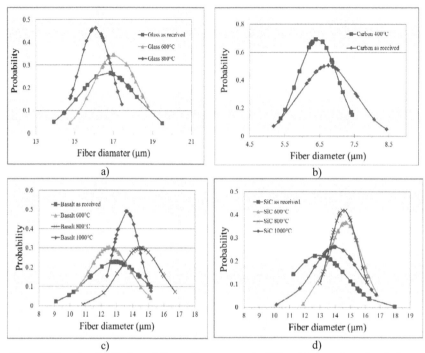

Figure 3 Diameter distribution of fibers after exposure to different temperatures for a) glass, b) carbon, c) basalt and d) SiC fibers.

The mean and standar deviation of the fiber diamter for the fibers at different oxidation temperatures has been included at Table 1. These mean values were used to calculate the tensile strength of the fibers.

Table 1 Main parameters obtained from the diameter measured for the fibers at the SEM.

Condition	Glass		Carbon		Basalt		SiC	
	Mean	SD	Mean	SD	Mean	SD	Mean	SD
As-received	16.671	1.497	6.740	0.789	12.876	1.747	14.466	1.570
400°C	16.811	1.241	6.438	0.573	12.702	1.233	14.523	1.673
600°C	17.083	1.317	-	-	12.544	1.153	14.679	1.088
800°C	16.054	0.862	-	-	14.472	1.336	14.563	0.950
1000°C	-	-	-	-	13.617	0.815	14.010	1.528

Figure 4 shows XRD for fibers after been exposed to different temperatures in an oxidation environment. It was found that glass fiber even though is melted after 600°C and then solidified, never became crystalline and its spectrum do not change, see Figure 4a.

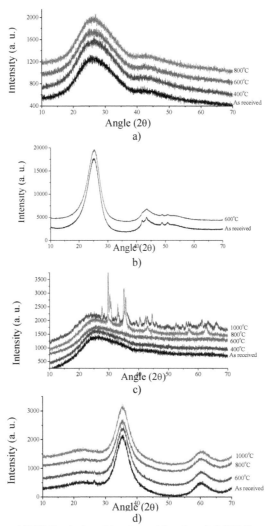

Figure 4 XRD for a) glass, b) carbon, c) basalt and d) SiC fibers after
exposure to different temperatures in air environment.

On the other hand, carbon fiber was only tested until 400°C because at 600°C it is almost
all gone into the air as byproducts of the oxidation (see Figure 4b). The XRD spectra at 400°C
and bellow are nearly the same and just show fiber remains as amorphous. For basalt fibers,
Figure 4c, after 600°C the fiber became chemically unstable since it starts changing from

amorphous to crystalline. This effect is reinforced after the 1000°C oxidation treatment. Finally, SiC fibers show the same XRD pattern during all conducted oxidation treatments.

Figure 5 shows the TGA data for different fibers from room temperature to 1000°C in air atmosphere.

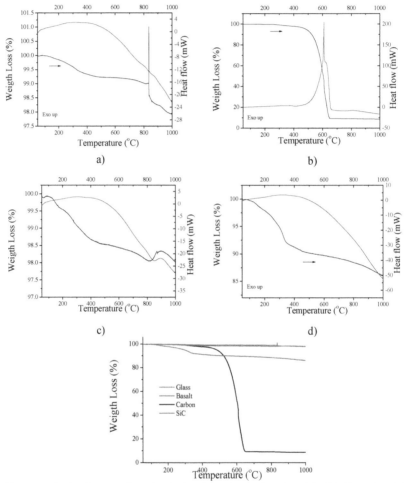

Figure 5 TGA in air atmosphere for a) glass, b) carbon, c) basalt and d) SiC fibers, e) weight loss in air atmosphere for all samples.

Figure 5a shows that glass fibers have very low weight loss over the full range of only just about 2%. The wide heat flow curve is due to the softening and further melting of the fibers which occurs almost over the full range but most significantly after 600°C and has a peak at 834°C due to an oxide formation. Figure 5b shows that carbon fibers have very significant transformations at 600°C. After 700°C the carbon is completely gone either as CO or CO_2, although this process started at about 400°C. Figure 5c shows that glass fibers have very low weight loss over the full range of only just about 2.5%. It is seen that there are several complex transformations that kind of corresponds to the changes already presented in the XRD. Figure 5d shows the results for SiC under air oxidation, which weight loss over the full temperature range is about 13%. Finally, all weight loss data have been summarized in Figure 5e. It is observed that glass and basalt fibers have very similar weight loss data, SiC fibers have more significant weight loss and carbon fibers are completely disappeared as a consequence of the oxidation.

Figure 6 shows the Weibull distribution for glass fibers after exposed to air oxidation at different temperatures for 1h. Fibers as received and after 400°C treatment do not show any significant difference from the tensile strength point of view. After 600°C treatment a strong deterioration due to softening and perhaps localized melting occurs and this is seen in the corresponding curve. For higher temperatures it was not possible to conduct any tensile experiments since the weakening of the fiber it was not possible to even prepare the sample for the test. It is important to note how variable the strength of these fibers is even in the as received condition, which is of about 1 order of magnitude.

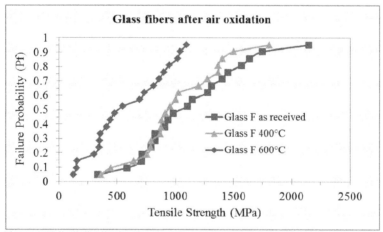

Figure 6 Weibull distribution for glass fibers after exposure to different temperatures.

Figure 7 shows the Weibull distribution for carbon fibers as received and after exposed to air oxidation at 400°C. At 600°C treatment a strong deterioration due to oxidation occurs (formation of CO or CO2) and this is seen in the corresponding curve the effect on the tensile properties. For higher temperatures it was not possible to conduct any tensile experiments since the fibers were completely gone into the air. Similarly as it was found for glass fibers, for carbon fibers the strength variability in all cases was very high, which is very important data for composites manufacturers.

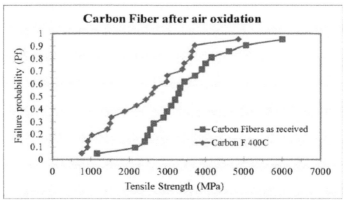

Figure 7 Weibull distribution for basalt fibers after exposure to different temperatures.

Figure 8 shows the Weibull distribution for basalt fibers after exposed to air oxidation at different temperatures for 1h. Fibers as received and after 400°C treatment do show a significant difference in the tensile strength. For fibers with the 600°C treatment there is a huge reduction in the tensile values, although the variability has been substantially decreased. At temperatures near 800°C and over it has been not possible to conduct any tensile experiments due to the weakening of the fiber.

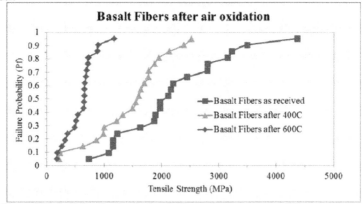

Figure 8 Weibull distribution for basalt fibers after exposure to different temperatures.

Figure 9 shows the Weibull distribution for SiC fibers after exposed to air oxidation at different temperatures for 1h. Fibers as received and after 400°C treatment do show a significant difference in the tensile strength. For fibers with the 600, 800 and 1000°C treatments there is a significant reduction in the tensile values and a reduce in the variability as well.

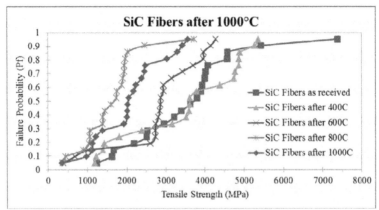

Figure 9 Weibull distribution for SiC fibers.

Figure 10 shows the Weibull modulus for the weibull distributions of the fibers presented above which is related to the variability in the tensile strength values. The values are low when they are compared to the same bulk materials or engineering ceramics.

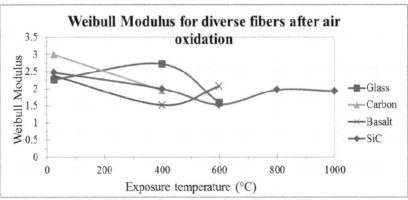

Figure 10 Weibull modulus for diverse fibers after exposure to different temperatures.

Figure 11 shows a summary of the mean tensile strength for the values presented before in the different Weibull distributions. In general it is seen that all fibers are deteriortaed by the air oxidation, even SiC after the oxidation at 400°C.

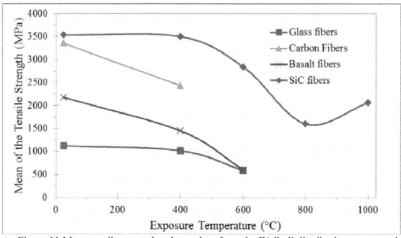

Figure 11 Mean tensile strength values taken from the Weibull distributions presented before.

CONCLUSION

Glass, carbon, basalt and SiC fibers have been oxidized in air for 1h at 200°C, 400°C, 600°C, 800°C, and 1000°C. In all fibers the tensile strenght was significantly affected by the temperature. Carbon fiber is comepletely destroyed after 600°C treatment by the oxidation. Only a little white ash remains. If temeprature is increased, this ash is also completely gone. Basalt fibers after the 600°C treatment are very weak and at 800°C they are partially melt and weak for which it is not possible to do tesnile tests. At 1000°C glass fibers are compeltely melted. SiC fibers do not melt, however after 1000°C their tensile properties are significantly decreased (more than 50%). In general it was found that all fibers were affected by the oxidation. A future communication will be reporting different solutions to protect the fiber from oxidation at tempeatures up to 1000°C including a method using chemically bonded ceramics[5-9] layers to protect the fibers.

ACKNOWLEDGEMENTS

The authors wish to thank to the NIST-ATP Program through a grant to Composites and Solutions Inc. (Program Monitor Dr. Felix H. Wu) and to Colciencias from Colombia for the grant to Henry A. Colorado.

REFERENCES
[1.] Feih, S., et al. "Strength degradation of glass fibers at high temperatures."Journal of materials science 44.2 (2009): 392-400.
[2.] Yin, Y., et al. "The oxidation behaviour of carbon fibres." Journal of materials science 29.8 (1994): 2250-2254.
[3.] Militký, Jiří, Vladimír Kovačič, and Vladimír Bajzík. "Mechanical properties of basalt filaments." FIBRES & TEXTILES in Eastern Europe 15.5-6 (2007): 64-65.

4. Varadachari, Chandrika, Ritabrata Bhowmick, and Kunal Ghosh. "Thermodynamics and Oxidation Behaviour of Crystalline Silicon Carbide (3C) with Atomic Oxygen and Ozone." ISRN Thermodynamics 2012 (2012).

5. Henry Colorado, Clem Hiel, H. Thomas Hahn and Jenn-Ming Yang Chemically Bonded Phosphate Ceramic Composites, Metal, Ceramic and Polymeric Composites for Various Uses, John Cuppoletti (Ed.), ISBN: 978-953-307-353-8, InTech, 265-282 (2011).

6. H. A. Colorado, C. Hiel, H. T. Hahn. (2011). Pultruded glass fiber-and pultruded carbon fiber-reinforced chemically bonded phosphate ceramics. Journal of Composite Materials, 45(23), 2391-2399.

7. H. A. Colorado. PhD thesis. Mechanical behavior and thermal stability of acid-base cements and composites fabricated at ambient temperature (2013).

8. H. A. Colorado, C. Hiel, H. T. Hahn. "Chemically Bonded Phosphate Ceramics with Different Fiber Reinforcements." Mechanical Properties and Performance of Engineering Ceramics and Composites VI: Ceramic Engineering and Science Proceedings, Volume 32: 181-188 (2011).

9. H. A. Colorado, C. Hiel, H. T. Hahn, J. M. Yang. "High Temperature Furnace Door Test for Wollastonite Based Chemically Bonded Phosphate Ceramics with Different Reinforcements."Processing and Properties of Advanced Ceramics and Composites IV, Volume 234: 269-274 (2012).

HEAT DIFFUSIVITY MEASUREMENTS ON CERAMIC FOAMS AND FIBERS WITH A LASER SPOT AND AN IR CAMERA

G. L. Vignoles[a], C. Lorrette[a,b], G. Bresson[a,c], R. Backov[d]

[a]University of Bordeaux, LCTS, 3 allée de la Boétie, F-33600 Pessac, France
[b]CEA, LCTS, 3 allée de la Boétie, F-33600 Pessac, France
[c]University of Bordeaux, I2M, 350 Cours de la Libération, F-33410 Talence, France
[d]University of Bordeaux, CRPP, 115 Avenue Schweitzer, F-33600 Pessac France

ABSTRACT

The heat diffusivity is rather easy to measure on material samples as soon as a flat sample of constant thickness can be produced; however this is not always possible. We will present and discuss two cases, both solved by the use of a laser spot, an IR camera, some image processing and data analysis. First, SiC foams were produced as small bulk samples without any clearly defined shape, excluding the possibility of using a rear-face recording of temperature evolution for the identification of heat diffusivity: methods based on the processing of front-face responses to a step impulse of a laser spot allowed the extraction of diffusivity and conductivity. Second, a bundle of isolated SiO_2 fibers has been examined with the same laser spot; a distinct data extraction technique has been set up and validated.

INTRODUCTION

Many advanced ceramics are intended for use in a high-temperature environment, and/or for heat transfer engineering. Accordingly, the knowledge of their thermophysical properties is of crucial importance. However, when developing a new ceramic-based material, one has frequently to face the situation in which the samples available for properties estimation are of reduced size, and/or of regular shape. This typical situation is of considerable annoyance when trying to measure a heat diffusivity coefficient, since the most classical method is the flash method[1], which requires flat samples with constant thickness. Another similarly difficult situation is the determination of fiber thermal properties, which is a key issue when designing composite materials with thermal and/or thermostructural applications. Many methods exist for carbon fibers, which are based on their rather elevated electrical conductivity[2] ; but not all of them are applicable to insulating materials such as glass or silica fibers. Photothermal microscopy[3] has been investigated on C/C composites[4] and in-situ characterization of a carbon fiber thermal diffusivity was made. Photoacoustics[5] and thermal flash under a microscope[6] have been used for the measurement of glass fiber heat diffusivity. Measurements in a periodic regime have been proposed for carbon and tungsten fibers at high temperatures, by using a laser source, a chopper, and an IR detector[7]. The thermal diffusivity was obtained from the evolution of the phase lag and of the amplitude as the distance from the excitation point increases.

We propose here an alternative, lightweight method, based on the use of a laser and an IR camera along with step excitation instead of a periodic signal. Though the excitation is the same on foams and fibers, the analysis of the response is very different. When the fibers are excited by the laser, they lose heat with respect to the surroundings; nonetheless, the heat loss coefficient can be identified together with the thermal diffusivity. On the other hand, for foams, the heat losses may be neglected safely and the heat diffusivity identification may be performed in several ways.

In the following sections, the experimental procedure and the employed models are given; then, various identification procedures are tested and compared.

EXPERIMENTAL

The studied foams are macro/mesocellular SiC or biosourced SiC/C foams prepared by a HIPE (High Internal Pressure Emulsion) route followed by chemical and thermal treatments[8,9]. The samples were placed on a substrate and were exposed to a 473 nm wavelength laser spot with 5 or 10 mW power, the laser direction making a 45° angle with the surface (see Figure 1). At 90° with respect to the laser direction, the foam heating was continuously recorded with an FLIR SC7000 IR camera with a cooled detector at 100 Hz or 760 Hz frequencies. The camera pixel size was recovered by in-situ calibration using the dimensions of the sample holder.

The characterized fibers are silica fibers with density 2.21 g.cm^{-3} extracted from a commercial fabric. The fiber bundle of diameter 0.45 mm is maintained by traction in a holder; it is painted in black, and it has been checked by scanning electron microscopy that the paint thickness was less than 0.3 μm, so that its impact on the effective thermal diffusivity is negligible. In order to minimize the system perturbation due to uncontrolled heat losses by natural or forced convection, the setup is placed in a vacuum cell, in which the pressure is lowered down to 10^{-3} Pa. The same exposure and recording procedure was then used with 10 mW power and 50 Hz acquisition frequency.

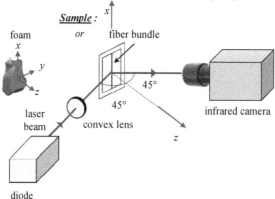

Figure1. Experimental setup for laser+IR camera heat diffusivity measurements.

The movies were then processed using FLIR proprietary software and ImageJ freeware[10]: frame averaging, computation and subtraction of the background, and image smoothing. Time evolution of the temperature on selected points as well as thermal profiles at selected instants, have been extracted for identification of thermophysical data by using the models presented in the next sections.

FOAMS: MODEL AND IDENTIFICATION PROCEDURES
The resolution of the heat equation:

$$\frac{1}{\alpha}\frac{\partial T}{\partial t} = \frac{\partial^2 T}{\partial x^2} + \frac{\partial^2 T}{\partial y^2} + \frac{\partial^2 T}{\partial z^2} \tag{1}$$

where $\alpha = \lambda/(\rho c_p)$ is the heat diffusivity (m^{2}s^{-1}), on a semi-infinite medium ($z \geq 0$) submitted to a continuous point energy source at (0;0;0) provides an adequate model to represent the temperature field evolution at the sample surface. Neglecting convective heat losses (i.e. $\frac{\partial T}{\partial z} = 0$ at $z = 0$), the surface temperature evolves with time according to the following formula[11]:

$$\theta(r,t) = \frac{q}{4\pi\lambda r}\,erfc\left(\frac{r}{\sqrt{4\alpha t}}\right) \tag{2}$$

where $\theta(r,t) = T(r,t) - T_0$ is the temperature in excess above the ambient (K), q is the heating flux (W.m^{-2}), r is the surface radial coordinate $\sqrt{x^2+y^2}$, λ is the heat conductivity (W m^{-1} K^{-1}), and $erfc$ is the complementary error function. This suggests two ways for thermophysical property identification. First, in the limit of r^2/t tending to large values, the $erfc$ term tends to 1, allowing identification of the thermal conductivity by fitting $q/4\pi\theta$ versus r. Dividing by the heat capacity (obtained $e.g.$ by differential scanning calorimetry) yields the desired quantity. However there are cases in which q is uneasy to estimate, since it is the ratio of the total power to the exposed surface (not precisely defined) times the emissivity (not precisely known). In that case, a nonlinear curve fit may be attempted by plotting $r\theta(r,t)$ versus $r/\sqrt{t}$ and fitting by an $erfc$ function. The scale factor of the $erfc$ argument gives the desired quantity. In addition to this method, identification of the thermal diffusivity is also possible by correlating the time derivative of the temperature to its apparent 2D Laplacian, without knowledge of the laser power or calibration of the IR camera signal.[12] During the test, there exist time and position values for which, at the surface, the z-directional term of the Laplacian $\dfrac{\partial^2 T}{\partial z^2}$ is negligible with respect to the 2D Laplacian $\dfrac{\partial^2 T}{\partial x^2} + \dfrac{\partial^2 T}{\partial y^2}$, that can be measured by image processing of the surface signal. When this occurs, the heat diffusivity is retrieved by plotting the time derivative with respect to the 2D Laplacian and taking the slope of the linear part.

FIBERS: MODEL AND IDENTIFICATION PROCEDURES

An approximate model of the experiment may be conveniently derived and exploited. The hypotheses of the model are: (i) 1D geometry; (ii) a constant heat flux q is injected at the spot center, considered as a single point; (iii) the bundle is considered symmetrical with respect to the spot; (iv) heat losses on the wire are considered as having a constant transfer coefficient h; (v) conductivity λ and heat capacity c_p are considered constant. The heat equation (1) is rewritten in 1D, integrating the heat losses in the balance:

$$\frac{1}{\alpha}\frac{\partial\theta}{\partial t} = \frac{\partial^2\theta}{\partial x^2} - \frac{4h}{\rho c_p D}\theta \tag{3}$$

where D is the fiber bundle diameter.

It may be fully solved in both cases of heating and cooling. We have obtained the solutions through the use of Laplace transforms[13]. For heating, one has:

$$\theta(x,t) = \frac{q\sqrt{\alpha/\beta}}{2\lambda}\left[\exp\left(-\sqrt{\beta x^2/\alpha}\right)erfc\left(\sqrt{\beta x^2/\alpha} - \sqrt{\beta t}\right) - \exp\left(\sqrt{\beta x^2/\alpha}\right)erfc\left(\sqrt{\beta x^2/\alpha} + \sqrt{\beta t}\right)\right] \tag{4}$$

where $\beta = 4h/(\rho c_p D)$ is the scaled heat loss coefficient.

Several identification methods appear, as discussed below. First, considering sufficiently large x while keeping low enough time values (so that $t \ll x/(2\sqrt{\alpha\beta})$), it can be shown that:

$$\frac{\partial\ln\theta}{\partial(x^2)} \approx -\frac{1}{4\alpha t} \tag{5}$$

Therefore, we may identify α by plotting the logarithm of the images $\ln\theta(x,t)$ versus x^2 at selected times and fit the curve with straight lines. Secondly, the steady-state solution of eq. (3) is:

$$\lim_{t\to\infty}\theta(x,t) = \frac{q\sqrt{\alpha/\beta}}{\lambda}\left[\exp\left(-\sqrt{\beta x^2/\alpha}\right)\right] \tag{6}$$

Consequently, the ratio $\sqrt{\beta/\alpha}$ may be immediately identified as the slope of $-\ln\theta = f(|x|)$. This only has an interest if we are able to identify independently the β coefficient. Indeed, this is the case, since it is possible to solve the heat equation for the cooling case. The solution is, for large enough values of t:

$$\ln\theta \simeq \text{const.} -\beta t - 1/2\ln(\pi\beta t) \tag{7}$$

The coefficient β may thus be identified by a plot of $\ln\theta = f(t)$ for any point chosen on the thermal profile. The thermal diffusivity may then be recovered from the values of $\sqrt{\beta/\alpha}$ in the steady-state and of β during cooling. The identification can be further refined by a full nonlinear least-squares fit of the experimental data $\theta(x,t)$ to eq. (4) during heating, giving improved values of α and β.

RESULTS AND DISCUSSION

The identification of the heat diffusivity from eq. (2) has been performed without trying to estimate the heat flux, by fitting the error function to $r\theta$ vs. $r/\sqrt{t}$ curves. For this, the spot location has been retrieved, and points were selected along the longest axis of the elliptical, concentric isotherms (do not forget that the images were acquired at $45°$ angle with respect to the surface normal). Figure 2 illustrates results obtained for one of the Bio-SiC/C(HIPE) foams, that are extremely diffusive, with a high sampling rate (760 Hz). Since all data points have been collected in less than one second, it has been verified that the heat losses are negligible for all considered values of $r/\sqrt{t}$. The identified diffusivities are very high (between 50 and 110 mm^2.s^{-1}); this validates the use of high acquisition frequencies.

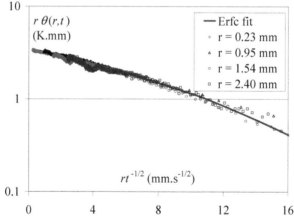

Figure 2. Bio-SiC/C(HIPE) foam characterization. Example of transformed time data for several points lying apart from the spot, and complementary error function curve fitting with $\alpha = 52$ mm^2.s^{-1}.

It has been found that this method may fail (i) because of a finite-size effect if the sample is too small and the diffusivity too high (indeed eq (2) applies strictly only to infinite media) or (ii) if the heat losses may become too important in the case of lower diffusivities. In the latter case, the time-gradient/Laplacian correlation method[11] could apply. To implement this method, image processing operations have been performed. First, the image is corrected for projections, so that the spot image becomes circular. Then, the 2D Laplacian operator has been obtained at every point for each frame by convolution with a 5×5 standard mask; the time difference is obtained by direct frame subtraction. All quantities have been smoothed with respect to time, and averaged along a radial profile taken along the longest axis of the elliptical isotherms, excluding the region of the spot itself. The 2D Laplacian is plotted as a function of the time derivative; when a linear relation is obtained, the proportionality constant (i.e. the curve slope) is taken as the thermal diffusivity. Tracking the correlation between the time evolution and the Laplacian, we find curves with two neatly distinct slopes (see Figure 3). For low values of the 2D Laplacian, a 2D heat equation approximates the material behavior well, and we can capture the diffusivity, which is between 31 and 7 mm^2 s^{-1}.

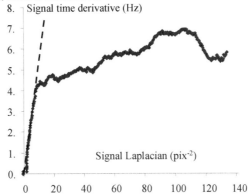

Figure 3. Identification of a heat diffusivity by the time derivative/Laplacian correlation method on an SiC(HIPE) foam. Here, 1 pixel is 0.155 mm, and $\alpha = 31$ mm^2.s^{-1}.

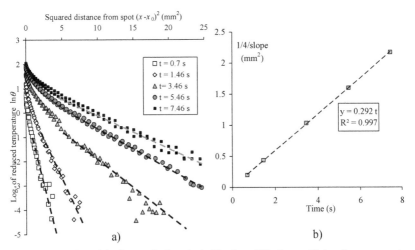

Figure 4. Illustration of the first method to obtain fiber heat diffusion coefficient from temperature profiles at selected times (a), then plotting the inverse of the slopes vs. time (b) and taking the slope of this trend.

Concerning the fiber characterization, the value of the diffusion coefficient was obtained by the three distinct methods mentioned in the preceding section, plus by the previously published periodic heating method[7]. Figure 4 illustrates the first method, for large enough x^2/t: the logarithm of the scaled temperature was plotted against the square of the distance from the spot; from the linear fits, the slopes are extracted (fig. 4a). The inverses of the slopes are plotted against time, giving the desired diffusion coefficient as a slope of this linear trend (Fig. 4b). The second method is illustrated at Figure 5. First, the ratio $\sqrt{\beta/\alpha}$ is extracted from the steady-state plot of $\ln\theta$, after shutting down the power, the cooling curve at any point (here, the hottest point, i.e. the spot location) is fitted at sufficiently large times with eq. (7), giving the β coefficient. The diffusivity value follows.

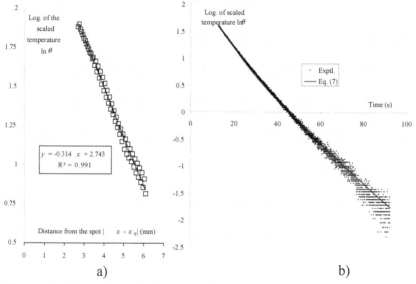

a)   b)

Figure 5. Illustration of the second method to obtain fiber bundle heat diffusion coefficient from (a) temperature profiles at steady state, and (b) fitting eq. (7) to a cooling curve at any point of the fiber bundle.

Last, it is even possible to fit directly the full solution for heating (eq. (4)) to the data set $\ln\theta(x,t)$. This has been conveniently done using the ad-hoc Levenberg-Marquhardt routine in the Gnuplot freeware[14], making use of the previous estimates as starting point for the minimization routine. Figure 6 illustrates the result. Measurements using the periodic heating method[7] have given identical values of the heat diffusion coefficient, within experimental error margins, as indicated in Table 1, where all results have been collected.

From this study, it appeared[13] that the first two methods are capable of larger uncertainties (1% instead of 0.01%) for several reasons. First, they do not involve the whole data set for identification of the coefficients. Second, the regions of the dataset in which the line or curve fittings are performed are not the regions where the signal/noise ratio is highest. Moreover, the second method rests on a measurement at steady-state: this is only attained approximately in the experiment. In addition, note that the exploitation of this experiment rests heavily on heat losses: we therefore have checked that it was necessary to maintain the heat loss coefficient as constant as possible, by enclosing the fiber bundle in a vacuum chamber with transparent walls.

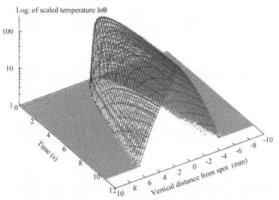

Figure 6. Full fit of eq. (4) for laser spot heating of a fiber bundle.

Method	$\alpha\,(\mathrm{mm}^2.\mathrm{s}^{-1})$	Error
1 : Large x	0.29	0.01 (3.5%)
2 : Steady+cooling	0.31	0.02 (7%)
3 : Direct fit	0.236	0.001 (0.5%)
Periodic	0.241	0.005 (2%)

Table 1. Summary of results on a glass fiber bundle.

CONCLUSION

We have presented a series of experiments for the determination of heat diffusivities in ceramic samples that are usually difficult to characterize because of their shape and size. Step heating with a single laser spot and an IR camera was used. Very high diffusivities have been measured on foams thanks to the large acquisition frequency of the camera and to its resolution; two methods allow coverage of at least the 10-100 mm^2.s^{-1} domain for foams. Fibers have been characterized using three distinct methods, two using only the heating data, and one requiring additional examination of the cooling behavior after laser shutdown. Values are in close agreement with each other and with a former method based on periodic heating. This method is practical since it does not require phase loop locking as in periodic methods, like the single pulse method[12]. The uncertainty due to the pulse duration in the latter method is avoided too.

Perspectives of this work address calorimetric data. Indeed, if the power of the laser is measured and if the sample dimensions are known, it becomes possible to push forward the exploitation of the presented model to the identification of the heat conductivity, and consequently of the heat capacity ρc_p as well as the effusivity $E = \sqrt{\lambda \rho c_p} = \rho c_p \sqrt{\alpha}$.

ACKNOWLEDGEMENTS
    This work was partly funded by GIS Advanced Materials in Aquitaine through a post-doc grant to G. B.

REFERENCES

[1] W. J. Parker, R. J. Jenkins, C. P. Butler, and G. L. Abbott, Flash Method of Determining Thermal Diffusivity, Heat Capacity, and Thermal Conductivity, *J. Appl. Phys.* **32**(9), 1679-1685 (1961).

[2] C. Pradère, J.-M. Goyhénèche, J.-C. Batsale, S. Dilhaire and R. Pailler, Thermal diffusivity measurements on a single fiber with microscale diameter at very high temperature, *Int. J. Thermal Sci.* **45**(5) 443-451 (2006)

[3] D. Rochais, G. Le Meur, V. Basini and G. Domingues, Microscopic thermal characterization of HTR particle layers, *Nucl. Eng. Des.* **238**(11) 3047-3059 (2008)

[4] J. Jumel, J.-C. Krapez, F. Lepoutre, F. Enguehard, D. Rochais, G. Neuer and M. Cataldi, Microscopic thermal characterization of C/C and C/C-SiC composites, in "Procs. 28th Quantitative Nondestructive Evaluation", D. O. Thompson, D. E. Chimenti, L. Poore, C. Nessa, S. Kallsen, eds., *AIP Conf. Procs.* **615**, pp. 1439-1446 (2002)

[5] D. Chardon D and S. J. Huard, Thermal diffusivity of optical fibers measured by photoacoustics, Appl. Phys. Lett. **41**(4) 341-342 (1982).

[6] M. T. Demko, Z. Dai, H. Yan, W. P. King, M. Cakmak and A. R. Abramson, Application of the thermal flash technique for low thermal diffusivity micro/nanofibers, *Rev. Sci. Instrum.* **80**(3) 036103 (2009)

[7] C. Pradère, J.-M. Goyhénèche, J.-C. Batsale, R. Pailler and S. Dilhaire, *Carbon* **47**(5) 737-743 (2009)

[8] S. Ungureanu, G. Sigaud, G. L. Vignoles, C. Lorrette, M. Birot, A. Derré, O. Babot, H. Deleuze, A. Soum, G. Pécastaings and R. Backov, Tough Silicon Carbide Macro/Mesocellular Crack-Free Monolithic Foams, *J. Mater. Chem.* **21**, 14732-14740 (2011).

[9] S. Ungureanu, G. Sigaud, G. L. Vignoles, C. Lorrette, M. Birot, H. Deleuze, and R. Backov Biosourced Syntheses of Monolithic Macroporous SiC/C Composite Foams: (Bio-SiC/C(HIPE)) and Study of Their Heat Transport Properties, *Chem. Mater.* , submitted (2012)

[10] http://rsb.info.nih.gov/ij/

[11] H. S. Carslaw and J. C. Jaeger, Conduction of Heat in Solids, 2nd ed., Oxford University Press, p. 261 (1959)

[12] V. Ayvazyan, J.-C. Batsale and C. Pradère, Simple possibilities of thermal diffusivity estimation for small-sized samples, with a laser pulse heating and infrared cameras, in *Procs. QIRT 2010*, X. P. V. Maldague, ed., Editions du CAO, Les Eboulements, QC, Canada, ref. 009 (2010).

[13] G. L. Vignoles, G. Bresson, C. Lorrette and A. Ahmadi-Sénichault, Measurement of the thermal diffusivity of a silica fiber bundle using a laser and an IR camera, in *Procs. Eurotherm 2012*, D. Petit & C. Le Niliot eds., *J. Phys.: Conf. Series* **395**, 012079 (2012)

[14] http://www.gnuplot.info

TOWARDS A MULTISCALE MODEL OF THERMALLY-INDUCED MICROCRACKING IN POROUS CERAMICS

Ray S. Fertig, III[a] and Seth Nickerson[b]
[a]Department of Mechanical Engineering, University of Wyoming, Laramie, WY, USA
[b]Corning Incorporated, Corning, NY, USA & Bauhaus University, Weimar, Germany

ABSTRACT
The tendency for ceramic materials with high thermal expansion anisotropy to crack at domain boundaries during thermal cycling has been extensively documented. However, much work still remains to develop a model that can predict this behavior as a function of thermal history. Here we present the development of a statistical model that accurately predicts the temperature history-dependent microcrack formation during thermal cycling of a porous ceramic. The functional form describing this mechanism was developed from a detailed study of a microscale finite element model, which is also discussed. Excellent agreement between experimental results and model predictions is shown.

INTRODUCTION AND BACKGROUND
Catalytic converters and diesel particulate filters are utilized in nearly every car or truck in the world to reduce harmful exhaust pollutants, such as nitrogen oxides (NOx), and to induce oxidation of hydrocarbons (HC) and carbon monoxide (CO)[1]. Since 1975, catalytic converter-equipped vehicles alone have helped cut air pollution by more than 3 billion tons worldwide, according to the Manufacturers of Emission Controls Association (MECA). Diesel particulate filters (DPF) physically prevent most harmful solid particulate matter produced by the engine, commonly referred to as soot, from entering the atmosphere. Both catalytic converters and DPFs use cellular ceramic substrates, which combine the large surface area of a honeycomb structure with a highly engineered ceramic material microstructure. This design has been so effective that the its developers were awarded the 2003 National Medal of Technology[2].

A key attribute of current honeycomb ceramic substrates used for emissions control is the ability to survive the extremely harsh operational environment of an engine's exhaust, which requires chemical stability and thermal shock resistance. Common ceramic materials such as mullite and alpha-alumina have coefficients of thermal expansion (CTE) 5-10 times higher than the commercial synthetic cordierite materials commonly used in the marketplace, rendering cordierite products much more thermally durable than many other materials for these applications. But although much is known about the unique mechanisms that give cordierite substrates their high thermal shock resistance, few material models exist that accurately describe the highly porous and heavily microcracked ceramic material behavior. The current work attempts to begin laying a foundation for a modular, multiscale material model to describe the non-linear temperature dependent stiffness properties of these materials.

Cordierite has a highly anisotropic crystalline CTE along its three crystalline axes[3-6] as shown in Figure 1. Remarkably, when embodied in the final honeycomb product the apparent CTE is much lower than one would expect of a randomly oriented polycrystalline material. This behavior is controlled by two primary mechanisms, preferential crystalline orientation and advantageous material microcracking[3-6]. Although these mechanisms are widely known[7,8], this work provides a mathematical framework to describe microcrack accumulation due to thermally induced micro-stresses and their effect on macroscopic material behavior.

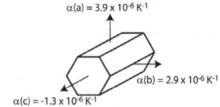

Figure 1. Thermal expansion anisotropy of cordierite

One of the most critical material properties is the elastic stiffness of the ceramic. However, the types of ceramics typically used in these applications are porous and may contain significant microcracking[4,9], which leads to properties that are not only dependent on temperature but also on microcrack density[10-12]. Typical behavior of the elastic modulus on cooling is shown in Figure 2 for a 50% porous cordierite honeycomb structure. The evolution of microcracking is dependent on the time-temperature history of the material—thus, the macroscopic material properties are not a state function of temperature[13-16].

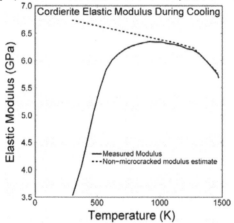

Figure 2. Cordierite elastic modulus during cooling

The focus of the work reported here is twofold. First, develop a finite element model to study the microscopic mechanisms of cracking in a ceramic to inform the development of a macroscopic model. Second, develop a macroscopic material model in light of any results revealed by the finite element model and experimental data observations.

FINITE ELEMENT MODELING

Model Setup
A two-dimensional finite element (FE) model was constructed in Abaqus[17] assuming hexagonal domains/grains, as shown in Figure 3 (left). Figure 3 (center) shows the mesh for a single grain, which was constructed of 4-node (linear) quadrilateral elements with incompatible modes (CPS4I). A layer of cohesive elements (COH2D4) was placed between each pair of hexagonal grains, as shown in Figure 3 (right). The size of the cohesive elements was 0.2 μm thick by 0.5 μm in length, much smaller than the elements in the grains, to avoid the well-

documented linkage between cohesive element size and measured fracture strength[18]. Tie constraints were imposed between all cohesive layer surfaces and the contacting grain.

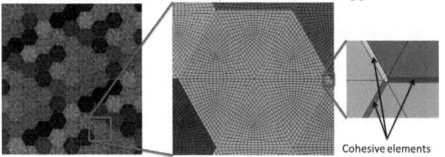

Cohesive elements

Figure 3. Meshing and element structure in FE model.

Representative crystalline cordierite material properties were used for the hexagonal grains. Each grain was assigned a random orientation based on a uniform random distribution between -90° and 90°. The anisotropic material properties, shown in Table 1, were taken published data. The stiffness and Poisson ratios were taken from Bubeck[6]. The anisotropic CTE values were taken from Bruno and Vogel[5].

Table 1. Cordierite Material Properties used in FEA model

Property	Value
Axial modulus (GPa)	148
Transverse modulus (GPa)	127
Shear modulus (GPa)	47
Poisson's ratio	0.34
CTE_{axial} (10^{-6}/K)	-1.2
$CTE_{transverse}$ (10^{-6}/K)	3.87

The constitutive behavior of the cohesive element is defined by a traction-separation law. This law describes two distinct regions of cohesive behavior: (i) an elastic region prior to failure initiation at some critical stress and (ii) a damage region where modulus degradation occurs with increasing displacement. The initiation criterion was a simple maximum stress criterion in both shear and normal stresses. A strain energy release rate of 0.573 J/m^2 was used to control element stiffness degradation.

For a traction-separation law, the stress formulation is based on the relative nodal displacements instead of strain. In the 2D case, the elastic constitutive law is completely defined by normal and shear spring stiffnesses, K_n and K_s, respectively. In order to ensure that the cohesive element does not add compliance to the model, K_i is typically chosen to be very large. Following the form proposed by Turon et al.[18], the stiffness values shown in Table 2 were chosen by dividing the appropriate cordierite elastic moduli by the thickness of the cohesive element (0.2 μm). For the studies reported here, the Abaqus-plugin Helius:MCT[19] was used to manage cohesive behavior convergence. This allowed viscosity issues typically associated with controlling cohesive behavior convergence to be ignored.

Table 2. Cohesive element stiffness

Cohesive properties	Value
K_n (GPa/m)	7.4×10^8
K_s (GPa/m)	2.35×10^8

Periodic boundary conditions (PBCs) on the edges of the model were considered most appropriate. Consequently, the model was constructed so that material orientations in each grain match at the boundary if the structure were tiled repeatedly. However, PBCs presented a particular problem when defining them on cohesive elements with a thickness of 0.2 μm, likely due to a penalty-based implementation of constraints. In addition, because nodes did not match between top and bottom boundaries, implementing these conditions rigorously would have required a great deal of additional effort to interpolate nodal displacements. As a compromise, the top and left edges were fixed in their respective normal directions. The bottom and right edges were constrained to deform uniformly in their respective normal directions. Temperature change ΔT was controlled using a homogeneous predefined field (isothermal), which was applied to the entire model and was varied to simulate a thermal cycle.

The goal of this FE study was to evaluate the effect of temperature change on mechanical behavior. Focus was particularly placed on the initiation of microcracks during cooling and corresponding decrease in elastic modulus. To continuously monitor the elastic modulus, the following method was employed. The displacement of the node at the bottom right corner of the model was tracked. This enabled extraction total strains in both the x- and y-directions during cool down. In addition, a small oscillatory load was applied to the node such that at every other increment the load was zero. During a loading increment, thermal strains were obtained via linear interpolation of the unloaded increments and an estimate of elastic strain from which the modulus was computed. However, during failure cascades linear interpolation of thermal strains is not accurate, and thus the significant numerical noise in the modulus can occur.

Finite Element Results

As a metric for comparison to experimental results, the evolution of elastic modulus was examined. Figure 4 shows the evolution of elastic modulus during a complete thermal cycle, where a stress free temperature of 1500K was assumed. Note in particular that no modulus recovery is observed during heating, in contrast to experimental observations. This is due to the fact that frictional effects, which aren't captured in the current model, are likely to be important and because healing is not incorporated in this model.

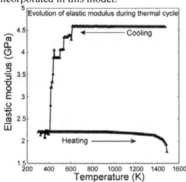

Figure 4. Evolution of elastic modulus during thermal cycling

One of the most advantageous features of simulating material behavior is the ability to query quantities that are difficult or impossible to experimentally observe. The stress in unfailed grain boundaries was one such quantity that was examined. Figure 5 (left) shows the distribution of boundary stresses for four different temperatures during cooling: 1165K, 625.2K, 448.4K, and 300.6K. (The starting temperature was assumed to be 1500K.) The primary features of importance are: (i) the stresses are distributed about a mean of roughly zero and (ii) the standard deviation increases with decreasing temperature. Furthermore, the stresses appear normally distributed, consistent with the results reported by Ortiz and Suresh[20]. The evolution of mean and standard deviation of boundary stresses are quantitatively characterized as a function of temperature in Figure 5 (right). We observe that the standard deviation increases linearly with decreasing temperature. The grain boundary shear stresses show the same trends.

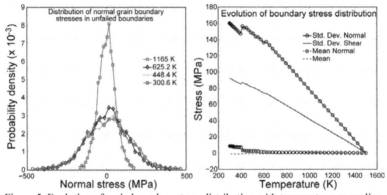

Figure 5. Evolution of grain boundary stress distribution with temperature on cooling

In addition to the stress distribution in the uncracked grain boundaries, the normal crack opening displacements were examined for a variety of temperatures on both heating and cooling. The probability densities for normal crack opening displacements at different temperatures are shown in Figure 6. The crack opening displacements range from nearly closed to nearly 300 nm. More results would need to be averaged to obtain the exact form of the distribution function describing these displacements, but it has roughly the appearance of an exponential distribution.

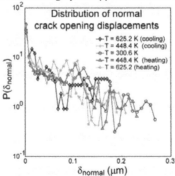

Figure 6. Distribution of normal crack opening displacement at various temperatures during heating and cooling

MACROSCOPIC MODEL

The majority of the change in the elastic modulus during cooling from a stress-free temperature is due to microcrack formation. During thermal cycling, microcracking can occur along the grain boundaries because of thermal expansion mismatch arising from anisotropy in the coefficient of thermal expansion (CTE).

The stress that drives microcracking at the grain boundaries can be estimated following Cleveland and Bradt[21] as

$$\tau = \frac{1}{2} E \Delta \alpha_{max} \left(T_0 - T \right) \qquad (1)$$

where E is the elastic modulus of the grain, $\Delta\alpha_{max}$ is the maximum difference in CTE between grains, and T_0 is the stress free temperature. In reality, because the orientation of the grains is represented by a distribution, $\Delta\alpha_{max}$ and, consequently, τ would be represented as distributions.

The focus of this work is to develop a model to predict microcrack formation during cooling. To describe microcracking in the ceramic at the macroscale we introduce a scalar damage variable n, which represents the density of microcracks N_C normalized by the total supply of boundaries that can crack N_S. This supply exists because only cracks with tensile normal forces can crack in our model. This effectively limits the number of grains that can actually crack, although this quantity is not known *a priori*.

$$n = \frac{N_C}{N_S}, \quad 0 \le n \le 1 \qquad (2)$$

A variety of equations have been proposed in published literature to describe the link between microcrack density with elastic modulus E^{10}. The functional form proposed by Hasselman and Singh[11] is used in this work to describe elastic modulus as a function of microcrack density as

$$\frac{E}{E_0} = \frac{1}{1 + \xi n} \qquad (3)$$

where ξ is a material specific parameter of order unity and E_0 is the uncracked modulus.

In Equation (3), the value of ξ was calculated assuming a fully microcracked material $(n = 1)$ at room temperature. The value of E_0 was approximated assuming bilinear temperature dependence. The calibration data for the model was a single cooling curve shown in Figure 2. The dashed curve represents an estimate of the behavior of the material with no microcracking E_0.

Here a new model is developed to predict microcracking that depends only on temperature change T^* from a zero-stress reference temperature T_0.

$$T^* = T_0 - T \qquad (4)$$

The finite element results revealed a stress distribution in the unfailed portion of the grain boundaries that was essentially Gaussian and that spread out with increasing T^*. This behavior is qualitatively shown in Figure 7, where S represents some strength of the grain boundary and τ is the grain boundary stress.

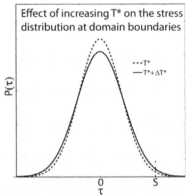

Figure 7. Effect of increasing T^* on the distribution of stresses at the grain boundaries

The rate of microcracking is proportional to the supply of cracks, given by $(1-n)$, and the rate at which the cumulative distribution between S and ∞ grows. For a mean stress of zero, which is consistent with our finite element results, the cumulative distribution for all stresses less than S is given by

$$C(S,\sigma) = \frac{1}{2}\left[1 + erf\left(\frac{S}{\sigma\sqrt{2}}\right)\right] \tag{5}$$

where erf is the error function and σ is the standard deviation of the stress τ. The cumulative distribution between S and ∞ is then given by $\left[1 - C(S,\sigma)\right]$. The finite element results (Figure 5 (left)) showed a linear relationship between standard deviation and T^* so that

$$\sigma = C_1 T^* \tag{6}$$

where, C_1 is a constant of proportionality. Using this relationship and defining the parameter

$$\varphi = \frac{S}{C_1\sqrt{2}} \tag{7}$$

allows the proportionality relationship for the rate of microcracking during cooling to be written as

$$\frac{dn}{dT^*} \propto (1-n)\frac{d\left(1 - C\left(T^*\right)\right)}{dT^*} \tag{8}$$

where $\dfrac{d\left(1 - C\left(T^*\right)\right)}{dT^*}$ is the rate at which the cumulative distribution increases. ϕ is treated as a constant in the equation such that

$$C\left(T^*\right) = \frac{1}{2}\left[1 + erf\left(\frac{\varphi}{T^*}\right)\right] \tag{9}$$

Differentiating Eq. (8) gives the rate of microcracking as

$$\frac{dn}{dT^*} = \beta(1-n)\frac{\varphi}{\left(T^*\right)^2}\exp\left(-\frac{\varphi^2}{\left(T^*\right)^2}\right) \tag{10}$$

where β and ϕ are the only adjustable parameters. The formulation of the model with respect to temperature and not time ensures temperature rate independence. In terms of time evolution, Eq. (10) can be written as

$$\frac{dn}{dt} = \beta(1-n)\frac{\varphi}{\left(T^*\right)^2}\exp\left(-\frac{\varphi^2}{\left(T^*\right)^2}\right)\left\langle\frac{dT^*}{dt}\right\rangle \tag{11}$$

where $\langle\ \rangle$ are Macaulay brackets to ensure that this mechanism only operates during cooling.

Figure 8 shows the fit between the model and the calibration data shown in Figure 2. The experimental open crack density was computed using Eq. (3). The fit is excellent, indicating the appropriateness of the functional form. This agreement is particularly remarkable given that the model was developed from a simple hexagonal grain structure and applied to a complex cordierite microstructure. Consequently, an important feature of Eq. (10) is that only two parameters are required, β and ϕ; effects of grain size, grain shape, porosity, pore morphology, grain boundary strength, and fracture toughness must be contained in these two parameters. The specific functional forms relating these parameters to microstructural features will be investigated in future work.

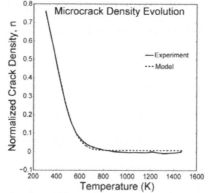

Figure 8. Comparison of the model with experimental microcrack evolution during cooling

Table 3 shows the model parameters derived from the calibration data for each mechanism. The model fit to the calibration data is shown in Figure 8. In general, the fit is very good. The slight difference at high temperatures may be due to the choice of stress-free temperature.

Table 3. Parameters fitted from calibration data

Parameter	Values
$\beta\ (K^{-1})$	645.8
$\phi\ (K)$	2522.3
$T_0\ (K)$	1554

SUMMARY AND CONCLUSION

In this work, we developed a finite element model to study thermally-induced grain boundary microcracking damage. This model was used to quantitatively evaluate the evolution of microcracking and grain boundary stresses during cooling from a stress-free temperature. The results obtained from the finite element model were used to inform the development of a macroscale model for microcracking. The realistic evolution of the elastic modulus suggest that even a geometrically simple model (hexagonal 2D grains) can be used for studying mechanisms of microcrack evolution.

The evolution of grain boundary stresses in the finite element model was used to develop the functional form for a statistical model describing microcracking in real cordierite microstructures. The resulting model was based on the cumulative distribution function of the grain boundary stresses relative to some critical grain boundary stress and showed that only two coefficients may be needed to describe the microcracking behavior. Thus, despite the simplicity of the finite element model and the complexity of a real cordierite microstructure, we showed that all of the microstructural complexity may result in essentially two terms that scale thermally-induced microcracking behavior.

To completely model thermal cycling behavior, a model is needed not only for crack initiation behavior described here, but also for crack closure, crack reopening, and crack healing. The crack opening distribution shown in Figure 6 is currently being used to develop a model for crack closure. Thus, the work reported here is only one component of what is required for a complete cyclic model. Nevertheless, the model is expected to accurately describe cracking in any temperature range of thermal cycling history where crack initiation is the dominant crack behavior.

ACKNOWLEDGEMENT

Financial support from Corning Incorporated via a subcontract to Firehole Composites is gratefully acknowledged.

REFERENCES
1   Favre, C., May, J. & Bosteels, D. EMISSIONS CONTROL TECHNOLOGIES TO MEET CURRENT AND FUTURE EUROPEAN VEHICLE EMISSIONS LEGISLATION. (Association for Emissions Control by Catalyst (AECC) AISBL, Brussels, 2011).
2   Corning, I. *Corning Research Team Awarded National Medal of Technology for the Invention of the Cellular Ceramic Substrate*, <http://www.corning.com/environmentaltechnologies/publications/ect_archives/ECT200 5-1-article4.aspx> (2012).
3   Heck, R. M., Farrauto, R. J. & Gulati, S. *Catalytic Air Pollution Control*. 3 edn, (John Wiley and Sons, Inc., 2009).
4   Bruno, G. *et al.* On the stress-free lattice expansion of porous cordierite. *Acta Materialia* **58**, 1994-2003, doi:DOI: 10.1016/j.actamat.2009.11.042 (2010).
5   Bruno, G. & Vogel, S. Calculation of the Average Coefficient of Thermal Expansion in Oriented Cordierite Polycrystals. *Journal of the American Ceramic Society* **91**, 2646-2652 (2008).
6   Bubeck, C. Direction dependent mechanical properties of extruded cordierite honeycombs. *Journal of the European Ceramic Society* **29**, 3113-3119, doi:10.1016/j.jeurceramsoc.2009.06.005 (2009).
7   Bush, E. A. & Hummel, F. A. High-Temperature Mechanical Properties of Ceramic Materials: I, Magnesium Dititanate. *Journal of the American Ceramic Society* **41**, 189-195 (1958).
8   Bush, E. A. & Hummel, F. A. High-Temperature Mechanical Properties of Ceramic Materials: II, Beta-Eucryptite. *Journal of the American Ceramic Society* **42**, 388-391 (1959).

9   Bruno, G. *et al.* Micro- and macroscopic thermal expansion of stabilized aluminum titanate. *Journal of the European Ceramic Society* **30**, 2555-2562, doi:DOI: 10.1016/j.jeurceramsoc.2010.04.038 (2010).

10   Zimmerman, R. W. The effect of microcracks on the elastic moduli of brittle materials. *Journal of Materials Science* **4**, 1457-1460 (1985).

11   Hasselman, D. P. H. & Singh, J. P. Analysis of Thermal Stress Resistance of Microcracked Brittle Ceramics. *Ceramic Bulletin* **58**, 856-860 (1979).

12   Feng, X. Q. & Yu, S. W. Estimate of effective elastic moduli with microcrack interaction effects. *Theoretical and Applied Fracture Mechanics* **34**, 225-233 (2000).

13   Shyam, A., Muth, J. & Lara-Curzio, E. Elastic properties of β-eucryptite in the glassy and microcracked crystalline states. *Acta Materialia* **60**, 5867-5876, doi:10.1016/j.actamat.2012.07.028 (2012).

14   Buessem, W. R., Thielke, N. R. & Sarakauskas, R. V. Thermal Expansion Hysteresis of Aluminum Titanate. *Ceramic Age* **60**, 38-40 (1952).

15   Shyam, A., Lara-Curzio, E., Pandey, A., Watkins, T. R. & More, K. L. The Thermal Expansion, Elastic and Fracture Properties of Porous Cordierite at Elevated Temperatures. *Journal of the American Ceramic Society* **95**, 1-10 (2012).

16   Bruno, G., Efremov, A., An, C., Wheaton, B. & Hughes, D. Connecting the macro and microstrain responses in technical porous ceramics. Part II: microcracking. *JOURNAL OF MATERIALS SCIENCE* **47**, 3674-3689, doi:10.1007/s10853-011-6216-y (2012).

17   Abaqus v. 6.11 (Dassault Systemes, 2011).

18   Turon, A., Davila, C. G., Camanho, P. P. & Costa, J. An engineering solution for mesh size effects in the simulation of delamination using cohesive zone models. *Engineering Fracture Mechanics* **74**, 1665-1682, doi:DOI: 10.1016/j.engfracmech.2006.08.025 (2007).

19   Helius:MCT v. 3.5 (Firehole Composites, Laramie, WY, 2010).

20   Ortiz, M. & Suresh, S. Statistical Properties of Residual Stresses and Intergranular Fracture in Ceramic Materials. *Journal of Applied Mechanics* **60**, 77-84 (1993).

21   Cleveland, J. J. & Bradt, R. C. Grain Size/Microcracking Relations for Pseudobrookite Oxides. *Journal of the American Ceramic Society* **61**, 478-481 (1978).

INVESTIGATION ON RELIABILITY OF HIGH ALUMINA REFRACTORIES

Wenjie Yuan, Qingyou Zhu, Jun Li, Chengji Deng, Hongxi Zhu
The State Key Laboratory Breeding Base of Refractories and Ceramics, Wuhan University of Science and Technology
Wuhan 430081, P.R. China

ABSTRACT

In this study, the flexural strength of high alumina refractories has been statistically analyzed by Weibull distribution based on the data of thirty-five bending tests for every kind of samples. The reliability of the flexural strength of high alumina refractories was evaluated. The phase composition and microstructure of samples was characterized by X-ray diffraction, chemical analysis and scanning electron microscope respectively. The results demonstrated that the difference of Weibull modulus value of high alumina bricks is obvious because of the different chemical, phase composition and microstructure. The ranges of failure probability were determined using the Weibull estimates.

INTRODUCTION

Refractories are key materials in high temperature industry including steel making, cement and glass manufacturing. Because of its poor plasticity, refractories are sensitive to several parameters including pores and phase compositions, and the random of the parameters leads to great scatter of strength data. Therefore the values of strength as well as their distribution or probability of failure have to be considered for design of metallurgical equipments. The quality and reliability of refractories are significant because increasing the service life could decrease the consumption of natural raw materials.

High alumina refractories are defined as those containing more than 45 percent alumina (Al_2O_3) produced by using calcined bauxite or fused and sintered alumina.[1] Because of its high thermal stability, refractoriness and slag resistance, high alumina refractories are widely used in the lining of steel-making furnaces, glass furnaces and cement rotary kilns. Alumina-silica bricks are manufactured on a tonnage scale, meaning that the products exhibit variability in properties attributable to this scale of manufacture. The differences between products with different qualities are usually not revealed in physical property tests alone.[2]

The Weibull distribution has been shown to be an appropriate model to describe strength data for ceramic materials.[3] The two-parameter Weibull model can be given by:

$$P = 1 - \exp\left[-\left(\frac{\sigma}{\sigma_0}\right)^m \right]$$

(1)

where P is the failure probability, m is Weibull modulus, σ is the flexural strength, σ_0 is the scale parameter or the Weibull characteristic strength corresponding to 63.2% failed.

The prediction of failure probability and reliability of refractories is a complex case. Based on the previous research, Weibull analysis could be used to evaluate strength of refractories.[4,5] In this work, the reliability of high alumina refractories was evaluated from the strength point of view. The flexural strengths of refractories including three kinds of high alumina bricks were measured. Weibull modulus of flexural strength was estimated and discussed combined with their apparent porosity, bulk density, chemical and phase composition and microstructure.

EXPERIMENTAL

Three kinds of high alumina refractories produced by a refractory factory were selected and numbered with 1#-3#. The chemical composition, apparent porosity and bulk density of three kinds of samples provided by the manufacturer are shown in Table 1. Each kind of high alumina bricks was shaped into 35 samples with the size of 25 mm×25 mm×150 mm. The surfaces of samples were polished. Three-point bending tests with the span of 100 mm were carried out at the loading speed of 0.15 MPa/s. The phase composition of bricks was determined by X-ray diffraction (XRD, X'Pert Pro, Philips). The amount of the phases was evaluated by the RIR method. The amount of amorphous phase and silica was measured by acid corrosion method.[6] Microstructure of samples embedded in epoxy resin after polished was observed by scanning electron microscope (SEM, Quanta 200, FEI).

Table 1. Chemical composition, apparent porosity and bulk density of high alumina bricks

| No. | Chemical composition (wt.%) | | | | | | Apparent porosity (%) | Bulk density (g/cm^3) |
	SiO$_2$	Al$_2$O$_3$	Fe$_2$O$_3$	CaO	MgO	Al/Si		
1#	30.80	65.61	1.13	0.28	0.38	2.51	37.97	1.77
2#	35.31	60.59	1.19	0.42	0.45	2.02	27.85	2.15
3#	35.35	60.74	1.11	0.38	0.40	2.02	26.49	2.22

RESULTS AND DISCUSSION

Flexural strength is listed with an order from small to large and shown in Fig. 1. The strengths determined from 1#-3# brick covered the range with 6.20 to 18.12 MPa, 5.12 to 14.33 MPa and 7.19 to 16.12 MPa, respectively. The results of Weibull statistic analysis are demonstrated in Table 2, from which it can be seen that the Weibull modulus and characteristic strength of 3# was larger than that of 1# and 2#. The larger Weibull modulus of 3# illustrated that the discrete of the distribution of strength for 3# was less. Weibull modulus of high alumina refractories is nearly or little less than that of most brittle ceramics ranging from 5 to 20.[7]

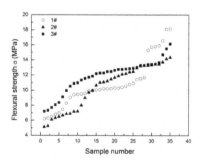

Figure 1. Distribution of flexural strength of high alumina brick samples

Table 2. Weibull modulus and characteristic strength of high alumina bricks

No.	Weibull modulus/ m	Scale parameter/σ_0	R^2
1#	4.06	11.94	0.8694
2#	3.79	11.57	0.9166
3#	5.98	12.79	0.9412

Probability density function $f(\sigma)$ is defined with the failure probability of materials under load range from σ to ($\sigma+\Delta\sigma$) in reliability engineering.[8] Reliability is defined with the unbroken probability of materials under load. Fig. 2 shows the relationship of flexural strength and probability density function. It can be seen that the probability density of high alumina refractory changed with its pressure increasing, and that curve was similar to the normal distribution form. There was the peak with higher f value and narrower width in the curve of 3# compared with others, which indicated that the failure strengths of 1# and 2# located in a wider range. The distribution pattern of probability density function presents a starboard type because Weibull moduli were both more than 3.6. The reliability was 1 under lower load, and the reliability decreased with increasing of load shown in Fig. 3. The reliability of 3# was obviously higher than 1# and 2# under the same load (less than 15 MPa).

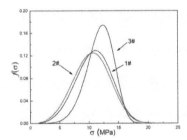

Figure 2. Relationship of flexural strength and probability density function

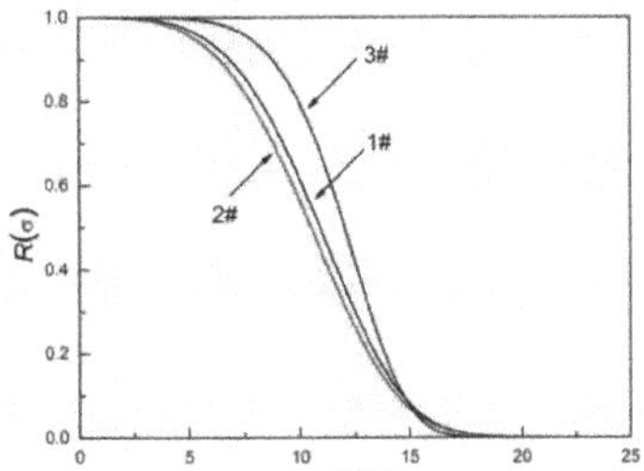

Figure 3. Relationship of flexural strength and reliability

The XRD patterns of high alumina bricks are shown in Fig. 4. In spite of alumina content in 1# was higher than others, the main peaks corresponded to corundum and cristobalite in the patterns of 1# brick. For 2# and 3# brick with the similar chemical composition, there was small different for the minor phases. According to the chemical composition listed in Table 1, Al/Si ratios of samples were less than stoichiometric ratio of mullite, that is to say, the excess silica would take part in formation of amorphous phase. Phase composition of high alumina bricks were calculated by the results of XRD and acid corrosion as listed in Table 3. Compared with the amount of phases, it was noticed that mullite content of 1# was less than others, moreover, amorphous phase of 1# was near twice as much as others. Though the amorphous phase contents of 2# and 3# were same, it could be deduced that the silica contents in the amorphous phase of 3# was less than that of 2# because the mullite and cristobalite contents of 3# were more than that of 2#. In one word, the content, composition and distribution of amorphous phase could determine the strength and its distribution of refractories.

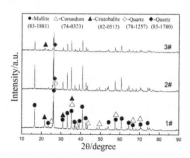

Figure 4. XRD patterns of high alumina bricks

Table 3. Phase composition of high alumina bricks (wt.%)

No.	Mullite (83-1881)	Corundum (74-0323)	Cristobalite (82-0512)	Quartz (78-1257)	Quartz (85-1780)	Amorphous phase
1#	63	16	2	—	—	19
2#	67	20	—	2	1	10
3#	70	16	1	—	3	10

From SEM images (secondary electron image) shown in Fig. 5, it can be seen that there were evident differences among microstructure of three kinds of bricks. Pores in 3# brick were much less than others. In addition, there were some cirques in cross-section of 3# brick. From these results, it can be presumed that alumina hollow balls may be used as raw materials.

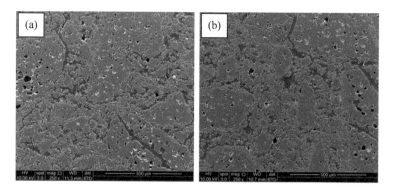

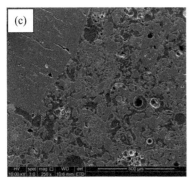

Figure 5. SEM images of high alumina bricks: (a) 1#, (b) 2# and (c) 3#

CONCLUSION

The results demonstrated that the flexural strength of high alumina refractories conforms to Weibull distribution. High alumina refractories with higher relative density and lower porosity are more reliable, and the discrete degree of the strength is less. Due to the difference of raw materials and processing, the amorphous phase in high alumina refractories has a remarkable influence on strength and its distribution. The reliability of flexural strength for high alumina refractories is determined by combined with chemical, phase composition and microstructure.

ACKNOWLEDGEMENT

This work was financially supported by National Key Basic Research Program of China (973) (No. 2012CB722702) and Key Project of Natural Science Foundation of Hubei Province of China (Grant No. 2010CDA023).

REFERENCES
[1]N. Li, H.Z. Gu, H.Z. Zhao, *Naihuo Cailiaoxue (Refractories)*, Metallurgical Industry Press, Beijing, 2010.
[2]C.A. Schacht, *Refractoires Handbook*, Marcel Dekker, Inc., New York, 2004.
[3]R.V. Curtis, A.S. Juszczyk, Analysis of strength data using two- and three-parameter Weibull models. *J. Mater. Sci.*, **33**, 1151-7 (1998)
[4]Q.Y. Zhu, W.J. Yuan, H.X. Zhu, C.J. Deng, Investigation on the bending strength distribution of MgO-C refractories. *Refract.*, **46**, 344-6 (2012)
[5]Q.Y. Zhu, W.J. Yuan, C.J. Deng, H.X. Zhu, Statistical analysis of the fractural strength for magnesia-chrome refractories via Weibull distribution. *Proceedings of The 6th International Symposium on Refractories*. Zhengzhou University Press, Zhengzhou, 508-11 (2012)
[6]J.P. Xu. The measurement of glass phase content in mullite. *Phys. Test. Chem. Anal. Part B: Chem. Anal.*, **43**, 605-6(2007)
[7]Y.W. Bao, Y.C. Zhou, H.B. Zhang, Investigation on reliability of nanolayer-grained Ti_3SiC_2 via Weibull statistics. *J. Mater. Sci.*, **42**, 4470-5(2007)
[8]Zwaag S van der, The Concept of Filament Strength and the Weibull Modulus. *J. Test Eval.*, **17**, 292-8(1989)

EVALUATION OF SUBCRITICAL CRACK GROWTH IN LOW TEMPERATURE CO–FIRED CERAMICS

Raul Bermejo[a], Peter Supancic[a], Clemens Krautgasser[b], Robert Danzer[a]

[a] Montanuniversität Leoben, Institut für Struktur- und Funktionskeramik, Leoben, Austria
[b] Materials Center Leoben Forschung GmbH, Leoben, Austria

ABSTRACT

Strength degradation in Low Temperature Co-fired Ceramics (LTCC) is strongly affected by the environment, causing slow propagation of cracks (SCCG) under stresses below the strength of the material. In this work the effect of humidity on the biaxial strength of bulk LTCC has been investigated using the ball-on-three-balls test (B3B) in different environments (i.e. water, air, silicone oil, and argon) at different stress rates (i.e. 0.01 MPa/s to 200 MPa/s). Whereas high strength values can be reached during high-rate testing in argon (inert strength), as low as 50% lower strength can be measured on specimens immersed in water tested for longer time periods. Constant stress rate experiments in a relative dry environment have shown for the first time evidence of two different crack growth mechanisms in LTCC material. A model to describe the crack growth rate (i.e. v–K curves) has been implemented in this work using a double power-law, which is based on the experimental strength results. The suitability of the B3B test for constant stress rate experiments is discussed and recommended for characterizing the SCCG in brittle materials and components.

KEYWORDS: Low Temperature Co-fired Ceramic, Environmental Assisted Cracking, Biaxial Strength, Inert Strength, Fracture Mechanics.

INTRODUCTION

Low-temperature co-fired ceramic (LTCC) is an important substrate material for the production of multilayer electronic circuits for medical, automotive and communication devices. The LTCC tape is produced by tape casting of a slurry consisting of low-melting-point glass particles, ceramic particles, a solvent, a dispersing agent, and an organic binder. The low sintering temperature of the ceramics in LTCC (around 900 °C) can be achieved by using a glass matrix with a low melting point, allowing a vitrification of the glass ceramic composite material [1; 2]. This makes the use of excellent electronic conductors such as silver, gold or mixtures of silver–palladium feasible, arranged within and/or on the surfaces of the ceramic substrate, forming complex multi-layered structures. Today, they can be found in electronic devices (e.g. for mobile and automotive technologies) which have to operate under harsh conditions such as relatively high temperatures and mechanical shock under different environments. Due to the different properties of the materials involved (e.g. thermal expansion coefficients, elastic constants, yield strength) LTCC components can be subject to internal stresses during fabrication, which may induce cracks that truncate the electrical performance of the device. It has been shown that metallization as well as surface features (such as contact pads or metal cavities) can affect the strength of LTCC components (see for instance [3-6]). Other processes after fabrication (e.g. packaging) can also involve high tensile stresses which may damage the component. In addition, mechanical stresses occurring during service conditions (e.g. rapid temperature changes, mechanical vibration, bending) can affect the structural integrity of the device.

A limiting factor for the lifetime of glass- and ceramic- based (brittle) components such as LTCC is associated with the subcritical crack growth (SCCG) phenomenon, especially in environments with high moisture content [7-13]. Existing cracks may grow under external applied stress even for values well below the characteristic strength of the material. In terms of linear elastic fracture mechanics, cracks may propagate (with a certain velocity, v) for an applied stress intensity factor, K, lower than the fracture toughness of the material, K_{Ic}.

For most ceramics and glasses, the basic crack growth rate, v, in the technically relevant velocity region (namely between 10^{-12} and 10^{-5} ms^{-1}), also defined as Region I (*reaction-rate controlled region*) is given by the empirical power-law relation [8; 10]:

$$v = \frac{da}{dt} = v_0 \cdot \left(\frac{K_I}{K_{Ic}} \right)^n \tag{1}$$

where a is crack size, t is time, n is the SCCG exponent and v_0 is a material constant (i.e. critical crack growth velocity) for a particular environment. Depending on crack growth rate and environment different mechanisms of SCCG may take place, which can be recognised by different slopes in a double log plot of the $v–K_I$ curve [8]. Many different ideas on mechanisms for SCCG can be found in the literature, which range from thermally assisted breaking of bonds at the crack tip [14], stress enhanced corrosion [9], hydrolysis of silica bonds to diffusion of water within the crack or the development of a diffusion zone around the crack tip which causes an increase of the fracture toughness [15]. In general the parameter n is used as a hint for the mechanism but a clear understanding of the physical principles of SCCG is still lacking [16]. Among the different models of crack growth, a direct chemical attack of the environment on the crack tip seems to have the strongest experimental support [17].

In order to obtain crack propagation data, both direct and indirect methods may be employed [9]. With direct methods crack velocity is measured on fracture mechanics type specimens (e.g. double cantilever specimen, double torsion specimen with a crack), as function of the stress intensity factor, by direct observation of the crack growth increment. With indirect methods the growth of internal defects causes a degradation of strength, which is used to derive the underlying crack propagation parameters. One indirect method easy to be performed is based on strength measurements at different stress rates in a given environment. Although only the average crack behaviour can be measured using this method, it allows direct testing of component–like specimens, so that extrapolation of strength data to real components is more accurate. A limitation of this method is associated with the inherent scatter of strength in ceramic-containing materials, related to the distribution of flaws (e.g. size and location) within the sample, as described by the Weibull theory [18; 19]. In this regard, testing methods to evaluate strength degradation due to SCCG should be very accurate to avoid measurements uncertainties [20], which can mask the scatter of strength as well as the effect of the environment.

In this paper, the mechanical strength of LTCC substrate was evaluated in different environments (i.e. with low and high relative humidity) to characterize the sub-critical crack growth behaviour in this material. Different loading conditions (i.e. from 0.01 to 200 MPa/s) and environments (i.e. water, air and silicone oil) were selected aiming to reproduce possible loading scenarios during service. Inert strength was measured in argon at a high stress rate. A biaxial testing method (the Ball-on-three-balls) was employed, which enabled accurate measurements of the mechanical strength of bulk LTCC samples. Based on the experimental strength results a model for the SCCG in LTCC was derived for low and high humidity environments using a double and a single power-law $v–K$ relation, respectively, which could characterize the crack propagation behaviour in different environments.

EXPERIMENTS

Material of Study

The substrate (glass-ceramic) employed to fabricate the LTCC specimens of study is made of approx. 50% of Al_2O_3 as filler and 50% of several glasses containing Ca, Na, Si, K, B and Al, where the crystallisation degree after sintering exceeds 90%. Microstructural characterisation of the bulk material can be found elsewhere [21]. For the biaxial strength measurements samples of "as-received" rectangular plate specimens of dimension *ca.* 11.0 x 9.6 x 0.41 mm^3 were used. For statistical significance, between 10 and 30 specimens were tested for each condition.

Experimental Evaluation of Biaxial Strength

The biaxial strength (maximum failure stress) was determined using the ball-on-three-balls test. The specimen was symmetrically supported by three balls on one side and loaded by the fourth on the opposite face (see Fig. 1). The loading balls had a diameter of 8 mm resulting in a support radius of 4.619 mm. A pre-load of 7 N was applied to hold the specimen in the fixture. More details on the testing methodology can be found in previous publications [22-25]. The tests were conducted under displacement control at different cross-head speeds (i.e. 0.0001 mm/min, 0.001 mm/min, 0.01 mm/min, 0.1 mm/min and 2 mm/min) using a universal testing machine (ZWICK Z010, Zwick/Roell, Ulm, Germany). The corresponding stress rates were calculated from the load-displacement curves, ranging from *approx.* 0.01 MPa/s to 200 MPa/s. The flexural strength was determined in every case from the maximum tensile stress (σ_{max}) in the specimen during loading, given by:

$$\sigma_{max} = f \cdot \frac{P}{d^2} \qquad (2)$$

with P [N] the maximum load at failure, d [mm] the plate thickness, and f a dimensionless factor which depends on the geometry of the specimen, the Poisson's ratio of the tested material and details of the load transfer from the jig into the specimen [24]. In order to determine the f factor a FEM analysis was performed using ANSYS 14.0 for this geometry [26], assuming isotropic elastic properties for the material, giving as a result: $f = 3.3 - 29.9 \cdot (d/L) + 187.8 \cdot (d/L)^2$. Where $L = 11$ mm is the width of the specimen. This factor f is valid for $0.4 < d < 0.5$ mm and $v = 0.2$.

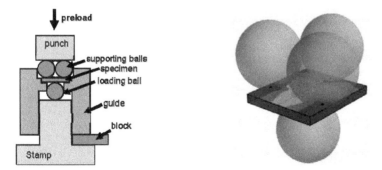

Figure 1. a) Schematic of the ball-on-three-balls testing jig; b) stress distribution in a plate during biaxial loading. The maximal stress is concentrated in the center of the specimen.

BIAXAL STRENGTH RESULTS IN DIFFERENT ENVIRONMENTS
The biaxial strength of LTCC specimens tested under several stressing rates (i.e. 0.01, 0.1, 1, 10 and 200 MPa/s) in different environments (*i.e.* silicone oil, air with 40 %RH, and water) was measured on samples containing between 10 and 30 specimens. The "inert" strength was attained in argon at high speed rate (200 MPa/s). As an example, the biaxial strength distribution in argon is represented in Fig. 2 in a Weibull diagram, where the failure stress, σ (calculated for every specimen according to Eq. (2)) is plotted versus the probability of failure, F. From the measured failure strength distribution the Weibull parameters, σ_0 and m, were calculated according to EN-843-5 [27]. The full line represents the best fit of the strength data using the maximum likelihood method. It can be observed that the failure stress values follow a Weibull distribution, which reflects the flaw size distribution in the sample [28]. All specimens failed at surface flaws located at the tensile surface. Fractographic evidences can be found elsewhere [21]. The characteristic flexural inert strength, σ_i, at room temperature resulted in 460 MPa with a 90 % confidence interval (CI) of $449 - 472$ MPa. The Weibull modulus, m, was 26 (relatively high for ceramic materials); the limits of the 90 % confidence interval ranged from 15 to 35. The relatively high Weibull modulus indicates a narrow critical flaw size distribution in the tested specimens. Moreover, it can also indicate the accuracy of the measurements (see Ref. [20] for more details).

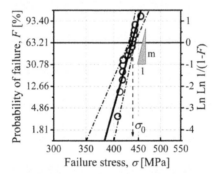

Figure 2. Weibull diagram of the inert strength of LTCC specimens tested in silicon oil at 200 MPa/s. The full line represents the best fit of the data using the maximum likelihood method and the broken lines show the limits of the 90 % confidence interval.

In Fig. 3 the strength data for each set is (log-log) plotted versus the stressing rate, tested in the three environments; the inert strength is also presented for comparison. The characteristic strength and Weibull modulus for each sample as well as the corresponding 90 % confidence intervals (error bars in Fig. 3) can be found in [21]. From Fig. 3 different slopes in the stress-stress rate curves can be seen, especially comparing humid environments (e.g. air or water) with dry conditions (i.e. silicone oil). For the cases of air and water, assuming a single power-law to describe the v-K behaviour, a SCCG parameter n ranging between 30 and 35 can be obtained [21]. However, the behaviour in silicone oil cannot be fitted using a single power-law. Based on the experimental evidences shown in Fig. 3, it can be observed how depending on the stress rate two different crack growth behaviours may be discerned. In order to understand the SCCG behaviour in dry environments (i.e. silicone oil in our case) a new model for crack growth must be implemented, and will be developed in the next section.

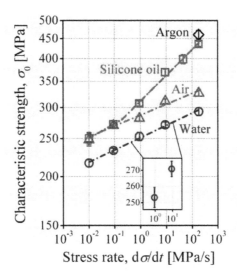

Figure 3. Characteristic strength as a function of the stressing rate (*i.e.* from 0.01 to 200 MPa/s) in different environments (*i.e.* oil (□), 40 %RH (△) and water (○)) presented as log σ_0 vs. log $d\sigma/dt$. The inert strength (◊) measured in argon is also shown as reference. The close-up shows the 90 % CI of strength of two samples measured in water.

MODEL FOR SUBCRITICAL CRACK GROWTH

Modelling the effect of SCCG by using a general crack growth law is based on the Griffith criterion, which predicts that fracture occurs when the stress intensity factor exceeds the fracture toughness of the material:

$$K_I \geq K_{Ic} \tag{3}$$

The stress intensity factor is usually given by:

$$K_I = Y \cdot \sigma \cdot \sqrt{\pi \cdot a} \tag{4}$$

while the critical stress intensity factor under inert conditions is defined by:

$$K_{Ic} = Y \cdot \sigma_i \cdot \sqrt{\pi \cdot a_i} \tag{5}$$

where σ_i is the inert strength and a_i the initial crack length[7].

---

[7] For this investigation, σ_i is set as 460 MPa and, accordingly, a_i calculated with Eq. (3) yields ~5 μm (assuming Y~1).

By introducing dimensionless quantities, $\hat{K}_I = K_I / K_{Ic}$ and $\hat{\sigma} = \sigma / \sigma_1$, a general evolution equation for the stress intensity factor can by derived:

$$\frac{d\hat{K}_I}{dt} = \frac{\partial \hat{K}_I}{\partial \hat{\sigma}} \times \frac{d\hat{\sigma}}{dt} + \frac{\partial \hat{K}_I}{\partial a} \times \frac{da}{dt} = \frac{\hat{K}_I}{\hat{\sigma}} \times \frac{d\hat{\sigma}}{dt} + \frac{\hat{\sigma}^2}{\hat{K}_I} \frac{1}{2a_1} \times \frac{da}{dt} \tag{6}$$

For constant stress rate experiments, as conducted in this investigation (i.e. $\dfrac{d\hat{\sigma}}{dt} = \dot{\sigma} / \sigma_1$), Eq. (6) reduces to:

$$\frac{d\hat{K}_I}{dt} = \frac{\hat{K}_I}{t} + \frac{1}{\hat{K}_I}\left(\frac{\dot{\sigma}}{\sigma_1}\right) t^2 \frac{1}{2a_{c,0}} \times \frac{da}{dt} \tag{7}$$

Substituting the crack growth rate associated with Region I, i.e. $da/dt = v_0(K_I/K_{Ic})^n$, into Eq. (7) we obtain a Bernoulli differential equation on $\hat{K}_I$, with t as independent variable:

$$\frac{d\hat{K}_I}{dt} = \frac{\hat{K}_I}{t} + \left(\frac{\dot{\sigma}}{\sigma_1}\right) t^2 \frac{v_0}{2a_{c,0}} \hat{K}_I^{n-1} \tag{8}$$

This differential equation can be solved even analytically in the case of a single power-law type crack growth behaviour, as derived in [29], resulting in:

$$\hat{K}_I = \frac{\dot{\sigma}}{\sigma_1} \cdot t \cdot \left(1 - \frac{n-2}{n+1} \cdot \frac{v_0}{2a_1} \cdot \left(\frac{\dot{\sigma}}{\sigma_1}\right)^n \cdot t^{n+1}\right)^{-1/(n-2)} \tag{9}$$

The time to fracture, i.e. $t=t_f$, can be calculated from Eq. (9) when $\hat{K}_I$ reaches 1. Then, for a given applied stress rate, $\dot{\sigma}$, the corresponding failure stress value is given by the applied stress at the fracture time, $\sigma = \dot{\sigma} \cdot t_f$.

Based on the experimental results in air and in water (i.e. characteristic strength vs. stress rate) Eq. (9) can be solved to best fit the parameters n and v_0 when fracture occurs (i.e. for $\hat{K}_I=1$), for different stress rates (i.e. taking $\dot{\sigma}$ as independent variable). The best fit is plotted in Fig. 4a along with the experimental strength data in both environments. The best fit for these data is obtained for $n\sim35$ and $v_0\sim0.5$ m/s for air and $n\sim31$ and $v_0\sim7$ m/s for water. Using these parameters, the v-K curve for LTCC can be obtained, as shown in Fig. 4b. It can be inferred from this figure that for a given K_I/K_{Ic} the crack growth rate in water may be more than 2 orders or magnitude compared to ambient air. A direct consequence is a significant reduction of the lifetime of LTCC material (see more details in [21]).

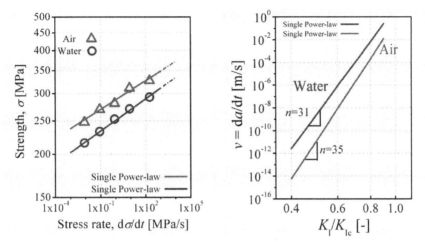

Figure 4. a) Characteristic strength as a function of the stressing rate (*i.e.* from 0.01 to 200 MPa/s) in water (○) and in air (△) presented as log σ_0 vs. log dσ/dt. The best fit is based on a single power law to describe the SCCG behaviour. b) v-K curve for water and air based on the parameters estimated from the strength-stress rate experiments.

In order to model the crack growth behaviour in silicone oil, a single power law could not account for the change of slope as found experimentally in the strength–stress rate tests (see Fig. 3). In an attempt to explain and rationalize the crack growth rate v as function of the applied stress intensity factor K in silicone oil, a double power law was implemented into Eq. (7) for da/dt, according to:

$$v(K_I) = \frac{da}{dt} = \begin{cases} v_0 \cdot \left(\dfrac{K_I}{K_{Ic}}\right)^n & K_I < K_T \\[2ex] v_1 \cdot \left(\dfrac{K_I}{K_{Ic}}\right)^{n_1} & K_I \geq K_T \end{cases} \tag{10}$$

Eq. (10) introduces five parameters into Eq. (9), i.e. v_0, n, v_1, n_1, and K_T, where four of them are independent parameters. The parameter v_1 can be expressed as: $v_1 = v_0 \cdot (K_T/K_{Ic})^{(n-n_1)}$. The parameter K_T is defined as "transition stress intensity factor" indicating the continuous transition between both crack growth regions, with different SCCG exponents, n and n_1, respectively. Now Eq. (7) can be solved in an analogous way and the evolution of $\hat{K}_I$ with time can be obtained. Solving for $\hat{K}_I = 1$, for different stress rates, the parameters v_0, n, v_1, n_1, and K_T can be found which best fit the experimental data in silicone oil. Since the region corresponding to low stress rates in oil and in air show the same trend (see Fig. 3) the parameters v_0, n, which describe Region I can be taken as those calculated for ambient air, i.e. n~35 and v_0~0.5 m/s. Thus, the best fit for oil is shown in Fig. 5a along with the experimental data. It can be seen how the model tends to a plateau region corresponding to the inert strength. However, this has not been measured with experiments yet.

Once the SCCG parameters for silicone oil have been obtained, the *v-K* curve (as given by Eq. (10)) for LTCC in silicone oil can be built up, as shown in Fig. 5b. The relationship between log strength vs. log stress rate leads to a kink which is related to the change in the crack growth mechanism occurring in the material during loading. At relatively low stress intensity factor, a *reaction–rate–controlled region* dominates the crack growth rate (defined as $v = da/dt$ [8]). At intermediate stress intensity factor the crack growth rate increases to a level limited by the rate of diffusion of water molecules to the crack tip, and thus a *transport–controlled region* governs the growth rate in the *v–K* curve [8]. The resulting parameters to describe SCCG of LTCC in oil may change in a different environment (e.g. argon) and temperature. This will be investigated in future work.

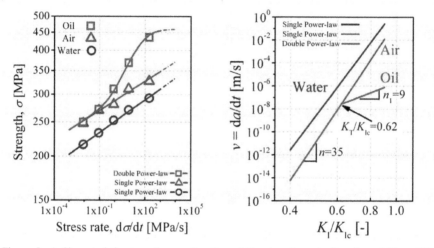

Figure 5. a) Characteristic strength as a function of the stressing rate (*i.e.* from 0.01 to 200 MPa/s) in water ($\circ$), in air ($\triangle$) and in silicone oil ($\square$), presented as log σ_0 vs. log $d\sigma/dt$. The best fit in oil is based on a double power law to describe the SCCG behaviour. b) *v-K* curve for water, air and oil based on the parameters estimated from the strength-stress rate experiments. A "kink" in the oil curve can be observed, indicating different SCCG mechanisms acting at the crack tip.

SUITABILITY OF THE BALL-ON-THREE-BALLS TEST
Several methods can be used to determine the biaxial strength of ceramic-based materials. Recent works on LTCC have employed the ball–on–ring (BOR) test [30] or the ring–on–ring (ROR) test [31] on bulk specimens, showing the effect of the loading rate and environmental conditions on the mechanical strength. These methods induce a maximal biaxial stress distribution in the centre of the specimen and avoid the influence of edge effects. However, it has been shown that, during testing, small geometric inaccuracies (e.g. for the case of "*as-sintered*" specimens) can lead to an undefined load transfer from the rings to the specimen and thus cause large uncertainties in the determined strength, as can be observed in [32]. Measurement uncertainties up to ± 10 % have been reported. This problem is specially enhanced when testing small specimens [33]. Other testing methods such as the pin–on–three–balls (POB) technique may

be used to avoid slight imperfections in the sintered plate [34], but there remain uncertainties about the loading under a flat-ended pin. Such uncertainties are much reduced (< 1 %) when a modified version of this loading configuration, the ball–on–three–balls (B3B) test, is used [22; 35-37]. Therefore, the effect of loading conditions, environment and/or surface features on the strength of such components may be better discerned using the B3B method, as has been reported elsewhere [4; 6; 38]. In Fig. 3 a close-up shows the 90 % CI of strength corresponding to two samples tested at different stress rates (i.e. at *approx.* 1 and 10 MPa/s). The narrow confidence interval for every testing condition represented by the scatter bars in Fig. 3 is not only a direct consequence of the high Weibull modulus of the material, but also is due to the accuracy of the testing method. In a previous work, it was demonstrated that a statistical error on the strength measurement of each specimen might mask the inherent scatter of the material strength [20]. Measuring the width of the strength distribution is only possible if the width of the distribution of measurement uncertainties (MU) is much smaller than the former. When methods such as ring-on-ring or ball-on-ring [39] are employed to determine the biaxial strength of brittle materials, MU up to 10 % may occur. They arise due to inaccurate positioning of the specimens, uneven contact and transfer of load, friction, etc. Such large MU are in the range of the scatter of the strength. It has been shown in [20] that, under these conditions, a Weibull modulus of about $m = 15$ would be determined, whereas the "real" modulus should be $m = 30$. In our experiments using the B3B technique, the experimental MU is less than 1 % and an accurate determination of the strength distribution (in particular of the Weibull modulus m) is possible. This testing method largely reduces uncertainties in the load transfer (e.g. due to lack of parallelism between both specimen sides) which might lead to under- or overestimation of the fracture load in the test.

In addition, the use of the B3B method, as inferred from Fig. 3, could allow discerning small differences in strength between samples tested in slightly different environments. Furthermore, as reported in this work, it was possible to observe the similar strength values in oil and in air for low stressing rates, which might have been mistaken by scatter using other testing procedures. In this work, more than 400 specimens were tested, which also demonstrates the ease and cost efficiency of the testing method. Thus, B3B is recommended for testing the SCCG behaviour of brittle materials.

SUMMARY AND CONCLUSIONS

In this work it has been shown that the biaxial strength of Low Temperature Co-fired Ceramic (LTCC) components is strongly affected by the environment and loading conditions in which the material is subject to mechanical stress. This is associated with the subcritical growth of cracks (SCCG) which is activated by the prolonged presence of humidity at the component surface loaded under tension. Whereas high strength values can be reached during high–rate testing in dry oil, up to a 50% lower strength can be measured on specimens immersed in water tested for longer time periods.

Constant stress rate experiments in a relative dry environment have shown for the first time evidence of two different crack growth mechanisms in LTCC material. A model to describe the crack growth rate (i.e. v–K curves) has been implemented in this work using a double power–law to interpret the experimental results. The model using a double power law can be used to describe the SCCG behaviour of other functional ceramic-based materials and can be useful to predict lifetime of components in dry environments.

The accuracy of the ball-on-three-balls method used for the biaxial strength measurements has enabled the effect of the stressing rate within a particular environment to be discerned, and is recommended for SCCG characterisation of glass containing ceramics as well as for other brittle materials.

ACKNOWLEDGEMENTS
Financial support by the Austrian Federal Government (in particular from the Bundesministerium für Verkehr, Innovation und Technologie and the Bundesministerium für Wirtschaft und Arbeit) and the Styrian Provincial Government, represented by Österreichische Forschungsförderungsgesellschaft mbH and by Steirische Wirtschaftsförderungsgesellschaft mbH, within the research activities of the K2 Competence Centre on "Integrated Research in Materials, Processing and Product Engineering", operated by the Materials Center Leoben Forschung GmbH in the framework of the Austrian COMET Competence Centre Programme, is gratefully acknowledged.

REFERENCES
[1] Y. Imanaka, Multilayered low temperature cofired ceramics (LTCC) technology, New York, NY 10013, USA, 2005.

[2] K. Makarovic, R. Bermejo, I. Kraleva, A. Bencan, M. Hrovat, J. Holc, B. Malic and M. Kosec, The Effect of Phase Composition on the Mechanical Properties of LTCC Material, *Int. J. Appl. Ceram. Techn,* **in press**, (2013).

[3] H. Dannheim, A. Roosen and U. Schmid, Effect of metallization on the lifetime prediction of mechanically stressed low-temperature co-fired ceramics multilayers, *J. Am. Ceram. Soc.,* **88**, 2188–94 (2005).

[4] R. Bermejo, I. Kraleva, M. Antoni, P. Supancic and R. Morrell, Influence of internal architectures on the fracture response of LTCC components, *Key Eng. Mat.,* **409**, 275-8 (2009).

[5] M. Deluca, R. Bermejo, M. Pletz, P. Supancic and R. Danzer, Strength and fracture analysis of silicon-based components for embedding, *J. Eur. Ceram. Soc.,* **31**, 549-58 (2011).

[6] R. Bermejo, P. Supancic, I. Kraleva, R. Morrell and R. Danzer, Strength reliability of 3D low temperature co-fired multilayer ceramics under biaxial loading, *J. Eur. Ceram. Soc.,* **31**, 745-53 (2011).

[7] C. Gurney, Delayed fracture in glass, *Proc. Phys. Soc. London,* **59**, 169-85 (1947).

[8] S. M. Wiederhorn, Influence of water vapour on crack propagation in soda-lime glass, *J. Am. Ceram. Soc.,* **50**, 407-14 (1967).

[9] S. M. Wiederhorn, Subcritical Crack Growth in Ceramics, *Fracture mechanics of ceramics,* Plenum, New York, 1974.

[10] A. G. Evans, Slow Crack Growth in Brittle Materials under Dynamic Loading Conditions, *Int. J. Fract.,* **10**, 251-9 (1974).

[11] R. Danzer, Ceramics: Mechanical Performance and Lifetime Prediction, *Encyclopedia of Materials Science and Engineering,* Pergamon Press, Oxford, 1993.

[12] EN 843-3, "Advanced Technical Ceramics, Monolithic Ceramics, Mechanical Properties at Room Temperature, Part 3: Determination of subcritical crack growth parameters from constant stressing rate flexural strength tests." In., 1996.

[13] R. Danzer, T. Lube, P. Supancic and R. Damani, Fracture of advanced ceramics, *Adv. Eng. Mat.,* **10**, 275-98 (2008).

[14] G. Schoeck, Thermally Activated Crack Propagation in Brittle Materials, *Int. J. Fract.,* **44**, 1-44 (1990).

[15] S. M. Wiederhorn, T. Fett, G. Rizzi, S. Fünfschilling, M. J. Hoffmann and J.-P. Guin, Effect of Water Penetration on the Strength and Toughness of Silica Glass, *J. Am. Ceram. Soc.,* **94**, s196-s203 (2011).

[16] S. W. Freiman, S. M. Wiederhorn and J. J. Mecholsky, Environmentally Enhanced Fracture of Glass: A Historical Perspective, *J. Am. Ceram. Soc.*, **92**, 1371-82 (2009).

[17] T. A. Michalske and S. W. Freiman, A Molecular Mechanism for Stress Corrosion in Vitreous Silica, *J. Am. Ceram. Soc.*, **66**, 284-8 (1983).

[18] R. Danzer, Some notes on the correlation between fracture and defect statistics: are Weibull statistics valid for very small specimens?, *J. Eur. Ceram. Soc.*, **26**, 3043-9 (2006).

[19] R. Danzer, T. Lube and P. Supancic, Monte Carlo simulations of strength distributions of brittle materials - Type of distribution, specimen and sample size, *Zeitschrift Fur Metallkunde*, **92**, 773-83 (2001).

[20] R. Bermejo, P. Supancic and R. Danzer, Influence of measurement uncertainties on the determination of the Weibull distribution, *J. Eur. Ceram. Soc.*, **32**, 251-5 (2012).

[21] R. Bermejo, P. Supancic, C. Krautgasser, R. Morrell and R. Danzer, Subcritical Crack Growth in Low Temperature Co-fired Ceramics under Biaxial Loading, *Eng. Fract. Mech.*, http://dx.doi.org/10.1016/j.engfracmech.2012.12.004, (2012).

[22] A. Börger, P. Supancic and R. Danzer, The ball on three balls test for strength testing of brittle discs: stress distribution in the disc, *J. Eur. Ceram. Soc.*, **22**, 1425-36 (2002).

[23] A. Börger, P. Supancic and R. Danzer, The ball on three balls test for strength testing of brittle discs: Part II: analysis of possible errors in the strength determination, *J. Eur. Ceram. Soc.*, **24**, 2917-28 (2004).

[24] R. Danzer, P. Supancic and W. Harrer, Biaxial Tensile Strength Test for Brittle Rectangular Plates, *J. Ceram. Soc. Jpn.*, **114**, 1054-60 (2006).

[25] R. Danzer, W. Harrer, P. Supancic, T. Lube, Z. Wang and A. Börger, The ball on three balls test - Strength and failure analysis of different materials, *J. Eur. Ceram. Soc.*, **27**, 1481-5 (2007).

[26] ANSYS, ANSYS Release 14.0, www.ansys.com (2011).

[27] EN 843-5, "Advanced Technical Ceramics - Monolithic Ceramics - Mechanical Tests at Room Temperature - Part 5: Statistical Analysis," pp. 40. In., 1997.

[28] R. Danzer, A general strength distribution function for brittle materials, *J. Eur. Ceram. Soc.*, **10**, 461-72 (1992).

[29] P. Supancic and H. Schöpf, Exact implementation of subcritical crack growth into a Weibullian strength distribution under constant stress rate conditions, *J. Eur. Ceram. Soc.*, **32**, 4031-40 (2012).

[30] H. Dannheim, U. Schmid and A. Roosen, Lifetime prediction for mechanically stressed low temperature co-fired ceramics, *J. Eur. Ceram. Soc.*, **24**, 2187–92 (2004).

[31] R. Tandon, C. S. Newton, S. L. Monroe, S. J. Glass and C. J. Roth, Sub-Critical Crack Growth Behavior of a Low-Temperature Co-Fired Ceramic, *J. Am. Ceram. Soc.*, **90**, 1527-33 (2007).

[32] R. Morrell, N. McCormick, J. Bevan, M. Lodeiro and J. Margetson, Biaxial disc flexure - modulus and strength testing, *Brit. Ceram. Proc.*, **59**, 31-44 (1999).

[33] T. Lube, M. Manner and R. Danzer, The Miniaturisation of the 4-Point Bend-Test, *Fatigue Fract. Engng. Mater. Struct.*, **20**, 1605-16 (1997).

[34] J. B. Wachtman, W. Capps and J. Mandel, Biaxial flexure tests of ceramic substrates, *J. Mater.*, **7**, 188-94 (1972).

[35] A. Börger, P. Supancic and R. Danzer, The Ball on three Balls Test for Strength Testing of Brittle Discs - Part II: Analysis of Possible Errors in the Strength Determination, *J. Eur. Ceram. Soc.*, **24**, 2917-28 (2004).

[36] R. Danzer, P. Supancic and W. Harrer, Biaxial Tensile Strength Test for Brittle Rectangular Plates, *J. Ceram. Soc. Jpn.*, **114**, 1054-60 (2006).

[37] R. Danzer, W. Harrer, P. Supancic, T. Lube, Z. Wang and A. Börger, The ball on three balls test – Strength and failure analysis of different materials, *J. Eur. Ceram. Soc.*, **27**, 1481-5 (2007).

[38] R. Bermejo, P. Supancic, F. Aldrian and R. Danzer, Experimental approach to assess the effect of metallization on the strength of functional ceramic components, *Scripta Mater.*, **66**, 546-9 (2012).

[39] R. Morrell, Biaxial flexural strength testing of ceramic materials. Measurement Good Practice Guide No. 12, Teddigton, UK, 1999.

MULTILAYER CERAMIC COMPOSITE ARMOR DESIGN AND IMPACT TESTS

Faruk Elaldi
Mechanical Engineering Department, Baskent University
Ankara, 06530 TURKEY

ABSTRACT
Alumina (Al_2O_3) ceramic with the properties of good hardness, thermal shock resistance, and chemical corrosion resistant features as aluminum based ceramic plate is sandwiched by two layers of Kevlar composite plates. The multilayer composite plates manufactured were tested by low velocity impact testing machine and damage analysis was made. Minimum impact energy level and its fragment-velocity equivalent were studied to be determined. The data obtained through impact testing were finally verified by Ansys /Autodyne FE analysis.

INTRODUCTION
There are several armor solutions for effective ballistic protection by using advanced steel alloys, ceramics and composite materials. One of these important solutions is the sandwich structure of alumina (Al_2O_3) ceramic together with Kevlar reinforced composite materials. Even though it requires high technology for the production of alumina and it is not easy to configure it for the complicated geometries, low density, good thermal shock resistance and higher hardness bring this material in front and make it a good candidate for armor solution [1-10].
In this study, locally produced alumina ceramic plates were sandwiched with 00/900 oriented Kevlar textile-epoxy composite layers from both sides. These combined armor plates were tested by low velocity impact testing machine and the data collected were evaluated for damage assessment considering perforation and penetration mechanisms. Thus, the minimum impact energy level that will not cause body injuring was tried to be determined. This attempt was also verified with a numerical method by using Autodyne finite element program.

MATERIAL AND FABRICATION
For the production of multilayer armor plates, Alumina was used as ceramic material and woven Kevlar fibers with areal density of 465 g/m^2 was used as reinforcement together with epoxy resin matrix. Three followings steps were applied to fabricate the armor plates.

Firstly, woven dry Kevlar were cut to the dimensions required for test specimens. In order to avoid being fuzzy at the cut-edges, cutting direction should be selected carefully and wetted by epoxy resin. For wetting, hand layup technique is applied using hand roller.

Secondly, alumina plates with a dimension of 30x40x2 in mm were placed without allowing any gap between plates on the epoxy wetted Kevlar layer by using a template of 100x100 mm, Figure 1 and 2.

Thirdly, the last layer of Kevlar/epoxy was transferred to the layup and ensured for wetting all over of the areas. The functionality of the outer layers of Kevlar/epoxy is to protect humans from alumina fragmentations after the panel gets hit and hold the alumina plates in proper positions during the curing processes. Later, overall layups were transferred to a hot press applying a pressure of 20 kg/cm^2 at a temperature of 140 ^0C. Test samples were cut to final dimensions of 100x100x4 mm after curing process was completed. Five identical test specimens were produced for low velocity impact testing and all armor test plates were x-rayed for inspection to be sure about proper ordering of alumina small plates in the sandwiched structure.

.

Figure1. Application of Kevlar/Epoxy Layers on Alumina Plates

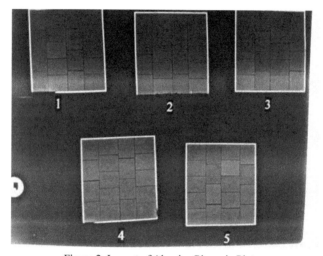

Figure 2. Layout of Alumina Pieces in Plates

EXPERIMENTAL STUDIES

Within the scope of this study, low velocity impact behavior of produced armor test plates is determined through a locally designed and manufactured instrumented impact test system, Figure 3. A drop-weight impactor with a spherical hardened steel head of 10 mm diameter was used. Test plates were fixed with upper and bottom templates in order to prevent flexure, Figure 4. The load, velocity, position and energy histories obtained from low velocity impact testing yield important information concerning damage initiation and growth. Plots of force, velocity, displacement and energy data provide impact characteristics of plates such as first failure, fracture forces, main damage formation, penetration, on-set of perforation and perforation limits. After test samples were first positioned in place of gripping area, the impactor is moved to touch the flush surface of the test samples in order to reset the position and load values. Later, the impactor are elevated to the required levels and released with free-fall to the surface of sample plates.

During the test, position and impact velocity data of the impactor are gathered from LVDT integrated to the testing device and force & energy data are collected from a dynamic load cell coupled with the impactor rod. All measuring data are fed to a data logger working together a signal conditioning unit and finally readings are displayed on a computer to be used for plotting graphics. All data collected are given in the following table, Table I.

Based on existing data collected, it is seen that sample-1 was totally penetrated, but other four samples were partially penetrated.

Figure 3. Low Velocity Impact Device and Test Sample-5

Table I. Experimental Results

Properties	Experimental Data					Numerical Data
	Test-1	Test-2	Test-3	Test-4	Test-5	
Mass (kg)	11.5	11.5	11.5	11.5	11.5	11.5
Velocity (m/s)	Unread	3.460	3.324	3.598	3.302	3.420
Height (mm)	600	600	570	600	600	600
Force (N)	Unread	107.915	106.674	163.724	136.742	-
Energy (Joule)	incalculable	68.83	63.53	74.43	62.69	67.25
Penetration	Full penetration	Partial Penetration	Partial Penetration	Partial Penetration	Partial Penetration	Partial Penetration

Figure 4. A View of Sample Gripping

MODELLING AND NUMERICAL ANALYSIS

In order to simulate the impact testing conditions; first, 3-D model was prepared, boundary conditions were determined and sample material specifications were found.

Samples were modeled by using CATIA/CAD program with some assumptions that all small alumina plates are touching each other without leaving any gap and epoxy film thickness used for adhesion of alumina plates is taken as an average value. Boundary conditions were selected according to the conditions that exist in testing device sample gripping area with upper and below fixing templates. The other important parameter to initiate correct numerical analysis is to use correct material specifications. It was preferred to use Autodyne library since we did not have alumina characteristics obtained from experimental tests. In the following Figure 5., the layers two and four from the top are representing adhesive epoxy film, one and five are representing Kevlar/epoxy composite layers and middle plane to represent ceramic alumina layer.

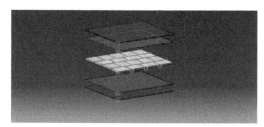

Figure 5. Alumina/Kevlar-Epoxy Sandwich Model

In this model, it was decided to use experimentally obtained average impact velocity, which was 3.42 m/s as the initial condition. On the other hand, 2 mm thick metallic templates used in fixing test samples were taken boundary conditions in the analysis, Figure 6. For simulating impact energy level obtained from the experimental tests, 22 caliber "long rifle bullet", Figure 7, was selected and used during the analysis.

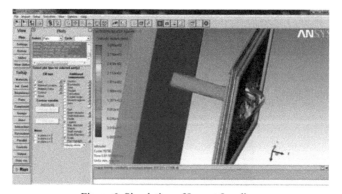

Figure 6. Simulation of Impact Loading

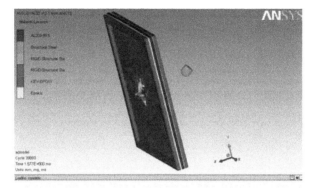

Figure 7. Penetration Simulation of 22 Caliber Long Rifle Bullet

The numerical analysis revealed that minimum energy level to partial penetration of the produced armor plates is roughly the same as the energy level obtained by the experiments. In other word, impact model works properly and the results verify the testing results, Figure 8.

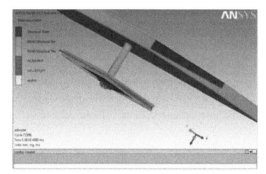

Figure 8. Simulation with Partial Penetration Energy Level

CONCLUSION

Numerical results and experimental results match with a negligible difference. The difference may come from using program library data for the material properties and possible gaps that may be occurred between small alumina plates during the fabrication phase of armor test samples.

The minimum energy level not to full penetrate the armor plates is found to be averagely 58.88 Joule by testing. That means that these ceramic armor plates are durable up to this energy level if they are ballistically impacted by a bullet or in other word, they will protect lives from catastrophic damages if the impact energy is not higher than that value. By comparison, this protection energy level is correspondent to the impact energy of 22 calibers bullet with a impact velocity of 225 m/s from 200 meters away from the target or the impact energy of 25 calibers machine gun bullet fired from 50 meters away. Thus, it is numerically proved that developed multilayer alumina/Kevlar composite armor plates provide full protection against to both aforementioned weapons.

ACKNOWLEDGEMENT
    Material support from NUROL TEKNOLOJI A.S., Ankara, Turkey is gratefully acknowledged. The support of Baskent University Mechanical Engineering Department is appreciated.

REFERENCES
[1]M. Übeyli, B. Ögel, Ballistic Performance of composites, Savtek, METU, Ankara, Turkey, 2002.

[2]B. Acar, Simulation of Composite Armors with Explicit Methods, Savtek METU, Ankara, Turkey, 2002.

[3]F. Şenel, L. Parnas, B. Balya, J. Garcia, Spall Liner on Armor Vehicles, Savtek, METU, Ankara, Turkey, 2006.

[4]N. Kol, R.O. Yıldırım, Reaktif Tablet Kullanılan Zırh Sistemlerinin Çukur İmla Jeti Delme Derinliğine Etkileri", Savtek, METU, Ankara, Turkey, 2008.

[5]S. Özel, R.O. Yıldırım, Patlayıcıyla Şekillendirilmiş Delicilerin Oluşumunda Astar Kalınlığının Delici Hızına Etkisi, Savtek, METU, Ankara, Turkey, 2008.

[6]G. Arslan, A. Kalemtaş, N.Tunçer, S. Yeşilay, F. Kara, S. Turan, Bor Karbür-Alüminyum Esaslı Zırh Sistemlerinin Balistik Performansının Belirlenmesi, Savtek, METU, Ankara, Turkey, 2008.

[7]B. Yayla, L. Parnas, F. Şenel, Ceramic Composite Armor Against Small Caliber (0-12,7mm) Bullets, Savtek, METU, Ankara, Turkey, 2006.

[8]Ş. Karagöz, A. Yilmaz, H. Atapek, Armor Steels and Developments, Savtek, METU, Ankara, Turkey, 2006.

[9]O. Ünal, Lecture Notes of Structural Materials, Afyon Kocatepe Üniversitesi 2011.

[10]P. Yayla, Fracture Mechanics, Çağlayan Kitabevi, 1. Baskı, 2007.

COMPRESSION FAILURE ANALYSIS OF GRAPHITE FOAM CORE BASED SANDWICH COMPOSITE CONSTRUCTIONS

Hooman Hosseini [a], Seyyed Reza Ghaffarian [a], Mohammad Teymouri [b], Ali Reza Moeini[b]

[a] Department of Polymer Engineering and Color Science, Amirkabir University of Technology, Tehran, Iran

[b] Petroleum Refining Development Division, Research Institute of Petroleum Industry, National Iranian Oil Company, Tehran, Iran

Abstract:

Coal tar pitch based graphite foams processed and examined under compression test in order to investigate the relation between microstructure and strength properties of microcellular graphite. Two step pyrolysis of coal tar pitch under pressure has been served to obtain microcellular carbonaceous solid which further carbonization converts it to the porous graphite structure with capability of usage as a core in sandwich structures instead of utilizing traditional honeycombs. After machining the as-obtained foams, compression test performed to investigate the mechanical properties of core. It was shown that by increasing the foaming pressure, compression strength and modulus increases. Amount of porous and formed microstructure within the solid has dominated effect in resistance to catastrophic failure. Increasing temperature in process would form the uniform open cell structure due to progressing in bubble nucleation and growth, which eventually shows higher ductility behavior during the collapse stage. Also it was observed that higher softening point of coal tar pitch and mesophase pitch precursors obtaining from first step heat treatment, enhances the compression strength due to increasing the melt strength and reaching higher molecular weight, leading more solidus matrix contains less air vacancies. Thus by altering the process parameters, mechanical properties of porous graphite material can be regulated.

INTRODUCTION

Lightweight sandwich structures are widely used in wind energy, aerospace constructions, marine, military artifacts and other high technological industries. Sandwich structures used in these applications, typically consist of a lightweight foam core bonded to thin face sheets to achieve high values of specific strength and stiffness.[1] Weight saving is another reason behind utilizing these structures instead of aluminum or polymeric bulky beams, One of the main advantages of sandwich structures is their potential to provide increased flexural rigidity without a considerable increase in structural weight.[2]

Basic element in weight reduction is the core material which is basically aluminum and paper-resin honeycombs or polymeric foams, giving a panel with enormous specific bending and compression stiffness and strength. Carbon foam core sandwich panels are novel products with exceptional properties. Carbon foams have been considered as a good candidate to be used as a core in structural elements within a sandwich construction since they could resist deformation perpendicular to the in-plane direction and provide shear rigidity along the planes perpendicular to the face sheets.[3] Furthermore, carbon foams can be joined using a wide range of commonly available adhesives to achieve large and intricate structures. Carbon foams can be readily engineered to meet particular component requirements. Their excellent machinability suggests that they are easy to cut, mill, turn, etc. using traditional machine shop equipment.

Development of these hybrid constructions requires full investigation of mechanical properties of carbon graphite foams. Innovation of carbon foam is attributed to the late 60's by introducing the vitreous glassy carbon foams.[4] Carbon foams have been produced from huge variety of precursors[5], nevertheless, pitch based carbon foams proved to have ultimate mechanical strength in comparison with phenolic or polymeric based carbon foams.[6] In addition, light weight features, conductive properties, fire and corrosion resistance introduced pitch based carbon foam as a brilliant high tech cellular product.[7]

In this work, carbon graphitic foams produced from two step heat treatment of coal tar pitch under pressure. In the first step, mesophase pitch has been synthesized from heat and pressure induced polymerization of raw coal tar pitch to get prepared for carbonization and graphitization in the second stage. Physical and mechanical properties of derived foam evaluated. Relative density and porosity as the basic structural features were calculated. The mechanical properties of carbon foams can be designed by changing its cell size, density, cell connectivity. After machining the as-obtained foams, compression test performed to investigate the mechanical properties of core. Effect of foaming pressure and soaking time during the bubbling process on the compression strength and modulus were investigated. Moreover, according to the stress-strain curves, failure mechanisms were evaluated. Also, effect of material parameters, like softening point of mesophase precursors on compression strength was studied. Considering the effects of these parameters, mechanical properties of porous graphite material can be regulated.

EXPERIMENTAL

Coal tar pitch prepared from Isfahan Coal Tar Refining Co. Specifications and elemental analysis of raw CTP presented in Table 1.

Table 1. Physico-chemical and Elemental properties of raw pitch.

SP[a]	TI[b]	QI[c]	CY[d]	AC[e]
105	33	12.5	54.6	0.2
C/N	N (%)	C (%)	S (%)	H (%)
120.1	0.776	93.23	0.359	6.669

[a] SP – softening point (°C)
[b] TI – toluene-insoluble content (%)
[c] QI – quinoline-insoluble content (%)
[d] CY – carbon yield (%)
[e] AC – ash content (%)
[f] D – density

All the heat-pressure treatments have been done in a cylindrical stainless steel reactor fitted with paddle type stirrer and equipped with temperature and pressure control unit. The reactor was optimized with the inlet and outlet inert purge of the nitrogen flow. Raw coal tar pitch converted to mesophase crystalline pitch using two process conditions as illustrated in Table 2. The reactor was heated at 3 °C/min up to the final temperature. During the heat treatment, an agitation of 200 rpm was maintained to assure the homogeneity in the isotropic and anisotropic formed sections and a flow of nitrogen 3 m^3/h was applied for the removal of volatile materials. The as-obtained mesophase pitches based on the Table 2 were named MP1 and MP2. Softening points were determined using ring and ball method, based on the ASTM D-36 standard procedure.

Table 2. Detailed process parameters of mesophase pitch production.

Sample Code	Precursor	Soaking time (hour)	Pressure (MPa)	Final Temperature (°C)	Softening Point (°C)	Density (gr/cm³)
MP1	CTP	6	5	420	171.5	1.325
MP2	CTP	2	-	400	192.5	1.431

In the second stage, MP1 and MP2 were crushed and sieved. Meso pitch powders were introduced into an aluminum mold having a predetermined shape and size. The reactor was washed with nitrogen to provide an inert atmosphere. The reactor was heated to 300 °C and at this specific temperature, different pressures, 2, 3, 4 MPa were applied, respectively, to obtain green foams with different bulk densities. Further heating raised the temperature to 450 °C. Soaking time of 3 and 6 hours applied at final temperature to obtain porous green coke. The as-received green foams were named as GF1-xy and GF2-xy, where x represents soaking time and y indicates the pressure used, and then the foams were carbonized and graphitized at 1200 °C for 1 hour in a purified nitrogen flow. (Table 3)

Table 3. Detailed process parameters of carbon foam production

Sample Code	Precursor	x (hour)	y (MPa)	Final Temperature (°C)
GF1-xy	MP1	3, 6	2, 3, 4	450
GF2-xy	MP2	3, 6	2, 3, 4	450

Density of graphite foams calculated based on the mass-volume method and porosity measurements conducted via Wang et al method [8] using suction apparatus and water absorption. Graphite foams were machined and subjected on the compression test according to the ASTM C365 (Figure 1). The size of cubic sample was $10 \times 10 \times 10$ mm. Stress- strain curves were plotted and compressive behavior of graphite foams was reviewed.

Figure 1. Load-cell apparatus for carbon foam compression measurement.

RESULTS AND DISCUSSION

Relative Density

Compressive strength of foams can be attributed to the foam structure parameters. The thickness of cell wall and the length of cell edge are used to describe the foam structure. The relationship among thickness of cell wall, length of cell edge, bulk density of specimens obeys from the Eq 1.

$$\left(\frac{\rho}{\rho_s}\right)^k \propto \left(\frac{t}{l}\right)^2 \qquad (1)$$

Where , ρ is the bulk density, ρ_s is the true density of the foam, and k is a constant between 1 and 2, t is the thickness of the cell wall and l is the length of the cell edge.[9] Cell wall thickness and cell edge length are the influential parameters in controlling the compressive strength of carbon foam.[10]

As figure 2 compiled, the average density of specimens GF1 and GF2 enhances with increasing pressure, which implies the thickening of cell walls and minimizing the cell edge length due to external pressure and subsequent packing in the system. GF1-64 represents the opposite trend. This fact can be attributed to the increasing the free volume, along of prolonging the soaking time from 3 to 6 hours for lower softening point pitch.

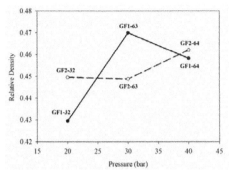

Figure 2. Relative density versus process pressure for GF1-xy and GF2-xy.

Porosity Measurement

Reduction of total porosity leads to the decrement in openings and free volumes and consequently increases the bulk density and close porosities inside the structure. This phenomenon enhances the strength of bulk by means of implementing less defects and open cell walls which are places for stress concentration and mechanical failures. Thus, as one can see, figure 3 depicts this trend for both low and high softening point precursor based GFs.

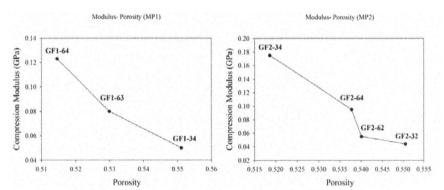

Figure 3. Relation of compressive modulus and porosity for GF1 and GF2 specimens.

Compressive strength of MP1-based carbon foams

Figure 4 depicts the compressive failure patterns of MP1-based carbon foams, according to S-S curves.

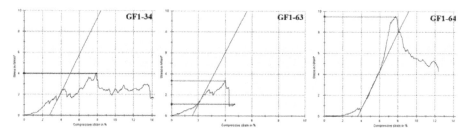

Figure 4. Load- strain curves for GF1 samples.

MP1 based foams show non homogenous elastic failure. This behavior was treated by increasing the foaming pressure and soaking time up to 4 MPa and 6 hours for GF1-64. Pressure augmentation leads to more uniform structure and dispersion, however, for GF1-64 density falls (Figure 2), which is attributed to increasing the free volume, nevertheless, one can notice that prolonging the soaking time in sufficient period to reach green coke stage with bulky graphite characteristics boosts the overall modulus of bulk graphite, regardless of opening development.

Overall, linear elastic region in compressive load-strain curves is related to bending of cell edges and stretching of cell faces. Non uniformity of this behavior is the cause of distortion of cells and initiation of cracks in cell walls, leading to distorted plateau region which is heterogeneous. Also, brittle crushing is the possible failure mechanism in GF1-34 and GF1-63. Collapse of foam cell would occur as a result of defects in cell walls and struts.

Comparison of GF1-34 and GF1-63 samples verifies the directive effects of relative density on mechanical properties. Densification is the major factor for stabilizing the compressive load and eliminating dramatic increase of strain. Densification leads to less initiation of crack and flaws and also distribution the pressure on every corner of porous pitch media to enhancing the compressive resistance.

It seems that despite of opening enhancement, increasing the soaking time and pressure would cause the densification of cellular structure and molecular weight of struts and ligaments by reaching to complete cellular coke artifact. So as one can see, using the maximum process parameters (4MPa and 6 hours) are more suitable for producing high strength carbon foam from low softening point precursors. Increasing the time, motivates the pyrolysis and poly condensation reactions which improves the layer growth and cross linking of aromatic planes in cell walls and bulk of porous material.

Compressive strength of MP2-based carbon foams

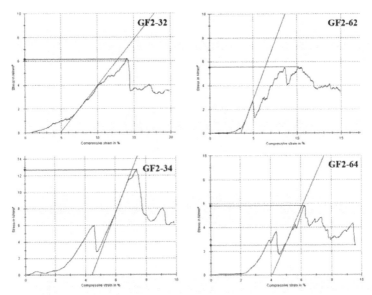

Figure 5. Load- strain curves for GF2 samples.

MP2-based foams show higher strength and modulus in more uniform trend. In more viscous media bubble can move slowly. So the possibility of collision and coalescence get weakened and the percentage of close pore spaces will become higher, consequently open spaces are less than GF1s and bulk strength is more than previous MP1-based specimens. Precursor with higher softening point has lower potential to be affected by pyrolysis conditions and removing of more volatile materials, thus pressure has dominated effect in controlling the structure and densification.

For GF2-32 and GF-62 samples with compression modulus of 0.044 GPa and 0.055 GPa, one can see positive effect of soaking time on modulus, due to decrease of strain and density. This fact is exceptional for MP2-based foams, however, it should be noticed that according figure 3, total porosity has dominating power in strengthen of compressive properties.

Failure mechanism of all MP2-based foams is elastic bending of cell edges with crack growth and propagation in cell walls, affecting by graphitization and delamination of graphene layers. Optimum conditions observed in GF2-34. 4MPa is enough to dominate over opposite impacts of soaking time in making puncture structure with poor mechanical ability, though; this resistance won't last for increasing soaking time to 6 hour. In the case of GF2- 64, existence of more orifice and interconnections as a result of coalescence and rupture of bubbles and pores weaken the load resistance.

CONCLUSION

Structural and mechanical properties of pitch based graphite foam as a high potential candidate in sandwich composite construction have been studied. Comprehensive evaluation of densitometry, porosity measurement and quasi-static compression are summarized in the following:

1. Higher softening point precursors have more potential to increase the stiffness and strength of resultant graphitic foam, due to higher viscosity of pitch media and constitution of higher molecular weight species.

2. Pyrolysis potency of pitch precursor is lower for GF2s and would lead the dominance of external pressure over soaking time. Therefore, utilizing 3 hours soaking time instead of 6 hours with 4MPa pressure produces optimum high strength graphite foam.

3. Conventional trend of total porosity and density versus external pressure have been observed. Densification and decrement of porosity were detected by pressure intensification. Partial deviations are observed due to pyrolysis and soaking time effects which are essential parameters for stabilization and carbonization of structure.

4. Failure mechanisms are mainly about crack initiation and growth in cell walls, cell walls lamination and cell face rupturing. MP2- based foams had uniform distribution of crack damping and higher crack resistance.

ACKNOWLEDGEMENT

This work has been done by the facilities and financial supports of Research Institute of Petroleum Industry (R.I.P.I.) in Iran and several fellow members in R.I.P.I-Bitumen and Road Construction Department.

REFERENCES
[1]Vinson J, Sandwich structures: past, present and future. *Netherlands*, Springer, 2005.
[2]Hosur MV, Mohammed AA, Zainuddin S, Jeelani S, Impact performance of nano phased foam core sandwich composites, *J Mater Sci Eng A*, **498**, 100–9 (2008).
[3]Hall R, Hager J, Performance limits for stiffness-critical graphitic foam structures. Part I: Comparisons with high-modulus foams, refractory alloys and graphite epoxy composites, *J Compos Mater*, **30**, 1922–37 (1996).
[4]Ford W, Method of Making Cellular Refractory Thermal Insulating Material, *US Patent 3121050*, (1964).
[5]Klett J.W, Mcmillan A.D, Gallego N.C, Walls C.A, The role of structure on the thermal properties of graphitic foams, *Journal of Materials Science*, **39**, 3659-3676 (2004).
[6]Klett J, Hardy R, Romine E, Walls C, Burchell T, High-thermal-conductivity, mesophase-pitch-derived carbon foams: effect of precursor on structure and properties, *Journal of Carbon*, **38**, 953-973 (2000).

[7]Reyes G, Rangaraj S, Fracture properties of high performance carbon foam sandwich structures, *Journal of Composites Part A: Applied Science and Manufacturing*, **42**, 1-7 (2011).

[8]Wang X, Zhong J, Wang Y, Yu M, Wang Y, The study on the formation of graphitic foam, *Journal of Materials Letters*, **61**, 741-746 (2007).

[9]Wun X, Fang M, Mei L, Luo B, Effect of final pyrolysis temperature on the mechanical and thermal properties of carbon foams reinforced by aluminosilicate, *J Materials Science & Engineering A*, **558**, 446–450 (2012).

[10]Gibson L.J, Ashby M.F, Cellular Solids- Structure and Properties, 2nd, *Cambridge University Press*, Cambridge, 1997.

Mathias Woydt[1], and Hardy Mohrbacher[2]
[1] BAM Federal Institute for Materials Research and Testing, DE-12200 Berlin, Germany
[2] Niobelcon BVBA, BE-2970 Schilde, Belgium

ABSTRACT
The tribological profile of alumina (99,7%) mated against binderless niobium carbide (NbC) rotating disks under under unlubricated (dry) friction and the type of motions of unidirectional sliding (0.1 m/s to 7.5 m/s; 22°C and 400°C) and oscillation (f= 20 Hz, Δx= 200 mm, 2/50/98% rel. humidity, n= $10^5/10^6$ cycles) was determined including the microstructure and mechanical properties. The obtained tribological data were benchmarked with different ceramics, cermets and thermally sprayed coatings. The established tribological profile revealed a strong position of NbC under tribological considerations and for closed tribosystems against traditional references, like WC, Cr_3C_2, (Ti,Mo)(C,N), etc..

INTRODUCTION
Transition metals of the groups IVB to VIB in the periodic table of elements have the potential to form extremely hard carbides or nitrides. Their most significant technical application until today has been reserved for tungsten carbide typically in form of cemented carbides. The applications of straight tungsten carbides range from metal cutting and wire drawing, as well as oil & gas drilling and mining, mineral and ground tools to rolling. For wear applications, hard metals are deposited as coatings by means of thermal spraying.

Significant increase in labor cost and taxes in China as well as in mining costs escalated the prices for tungsten. Increasing domestic consumption and export quota in China represent other factors.

Niobium carbide that has been well known for decades however remained on a low knowledge profile. The poor sintering ability may have represented one of the causes, which can today be overcome by either hot pressing, high-frequency induction heated sintering or by plasma-spark sintering. Ceramic composites containing niobium carbide suited for electrical discharge machining were elaborated.

Nevertheless, little data on the tribological profile of niobium carbide are available. The tribological properties of NbC are unexplored, although NbC is a property-determining constituent in many steels and hard metals as well as in cast irons[1,2].

Densification of pure niobium carbide by hot pressing enables to determine the fundamental contribution of NbC to the tribological profile, including tribo-oxidation, without having side effects from binder metals or sintering aids.

EXPERIMENTAL PROCEDURE
Niobium carbide was produced from a commercially available high purity niobium pentoxide (Nb_2O_5) powder (CBMM grade HP311). In this process lignite was blended with Nb_2O_5 powder and subsequently loaded into a furnace that was pre-heated to a temperature slightly above 1200°C under a mixture of 5%H_2/95%N_2 as operating atmosphere. The progress of the carburization reaction was monitored by measuring the CO(g)-concentration in the out-going gas. After completed carburization the sample material was moved into a cooling zone and held under a slightly reducing atmosphere consisting of N_2/H_2.

-Hot-pressing

Fraunhofer IKTS (Dresden, Germany) manufactured 18 disks with a diameter of 60 mm and a height of 6 mm. Hot-pressing was performed without the use of sintering additives at 2150°C under 50 MPa (4 h, 10 K/min) reaching an average density of 7.68 gr/cm³. Entirely cubic NbC corresponding to JCPDS 38-1364 was identified by XRD in the hot-pressed disks. The bending bars as well as the specimens for sliding and oscillating tests were machined from the disks by means of electrical discharge machining (EDM). The microstructure (see Figure 1) analyzed by FESEM (Field Emission SEM) revealed dense sintering along the NbC grain boundaries. The micro-hardness over the disk height and the diameter in the middle of the height was uniform and averaged to 1681 ± 92 HV0.2. The HP-NbC did however not reach the theoretical density, due to pre-existing porosity in the NbC grains of the synthesized powder. This grain porosity is also responsible for the reduced micro-hardness as compared to literature data.

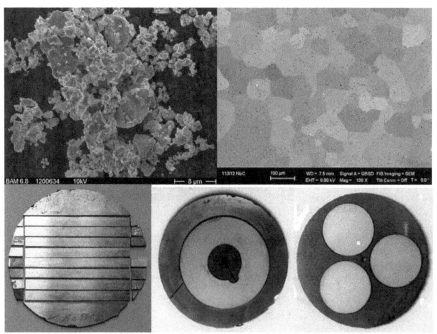

Figure 1. Morphology of powder (top left), FESEM of hot-pressed NbC and different samples ED; machined from disks (bottom raw)

-Mechanical properties

The four-point bending strength (3x4x45 mm) at 22.3°C (rel. humidity= 33%) was determined in a INSTRON machine equipped with 5000 N load cell meeting class 1 according to EN 10002-2 and by using a loading device (cross head speed= 0,8 mm/min) as required by DIN EN 843-1. The statistical Weibull analysis was performed according to DIN EN 843-5:2007 and is shown in Figure 2. It is obvious, that future grades without the intrinsic porosity in the grains will present significantly enhanced mechanical properties. The mechanical properties shown in Figure 2 are at least satisfactory for the determination of the tribological profile.

The elastic modulus was determined at room temperature on the same bars using the resonance method with a piezo-electric emitter-receiver in an ELASTOTRON 2000 machine. The elastic modulus calculated to 477 GPa using ASTM E1875-2008.

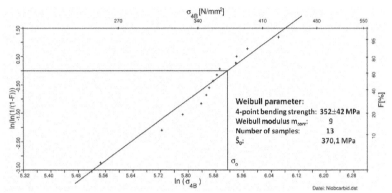

Figure 2. Weibull plot of σ_{4pb} at RT of HP-NbC1

-Tribometers

The tribometers for unidirectional sliding[3,4] and oscillating[5] sliding are proprietary developments of BAM and the details are disclosed elsewhere. They comply with ASTM G99 (DIN 50324) and with DIN EN 1071-13:2010. The wear volumes of stationary and rotating/oscillating specimen were calculated from stylus profilometry and the wear scar diameters by using ASTM D7755-11. The wear rate k_v is defined as the ratio of volumetric wear to the product of load F_n and the sliding distance s. The coefficient of friction (CoF) and the total linear wear of both tribo-elements (specimen) were recorded continuously. One test per combination of parameters was performed in this study.

--High-temperature tribometer

Sintered alumina (99,7%) bodies were used as stationary spherical (toroids with R_1 = 21 mm and R_2 = 21 mm) specimens with polished surfaces (R_{pk} = 0,019 μm), which were pressed against the planar surfaces of the rotating NbC specimens (lapping; R_{pk} = 0,1 μm). A normal force of 10 N was applied, resulting in an initial Hertzian pressure $P_{H\,0max}$ of approximately 660 MPa. The sliding distance was 5.000 m. Experiments were performed at 23°C and 400°C in air (rel. humidity at RT approx. 35%) with sliding speeds of 0.1, 0.3, 1.0, 3.0 and 7.5 m/s. The resolution limit of the wear rate for the rotating specimen corresponds to about 10^{-8} mm³/Nm.

--Oscillating tribometer

The oscillating tribometer uses a polished ball (∅=10 mm; alumina 99,7%) is fixed at the top of a lever with an integrated load cell for the measurement of friction force. The ball (not rotating) is positioned on the disk that is fixed on a table oscillating with 20 Hz and a stroke of 0.2 mm as well as is at 22°C loaded by a dead weight acting as the normal force (F_N= 10 N) perpendicular to the sliding direction. The tests were run under three relative humidity levels of 2%, 50% and 98% over one million of cycles. The sensitivity of a couple against humidity and the impact of tribo-oxidation can be effectively quantified under dry oscillation.

TRIBOLOGICAL RESULTS

The following tribological data under dry friction were compared with homologous results issued from the tribological data base TRIBOCOLLECT of thermally sprayed coatings[6], self-

mated ceramics, ceramic composites [7] and steels as well as mated with stationary specimen in alumina.
-Dry sliding

The frictional level in Figure 3 of NbC is as high as for different tungsten carbide based or Cr_2O_3–based hard metals or monolithic alumina and thus qualifies NbC for traction & frictional applications rather than for bearings. In comparison, HP-NbC comprised a particularly high wear resistance especially at RT and below 1 m/s sliding speed. The wear resistance of HP-NbC at RT is one of the lowest and compares well with that of self-mated alumina (Al_2O_3) couples also regarding its evolution with sliding speed.

At 400°C, the dry sliding wear resistances of tribo-active materials ($Ti_{n-2}Cr_2O_{2n-1}$-phases, (Ti,Mo)(C,N)), HP-NbC and thermally sprayed Cr_2O_3 or WC-based hard metals ranged between 10^{-6} mm^3/Nm to 10^{-7} mm^3/Nm on a level of mixed/boundary lubrication. Wear resistance of HP-NbC is better than that of Cr_3C_2 and better or similar than that of WC-based systems.

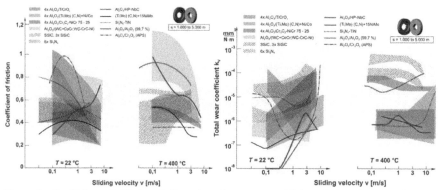

Figure 3. Coefficient of friction (left) and total wear rates (right) of HP-NbC and different ceramics and hard metals under dry friction for RT and 400°C

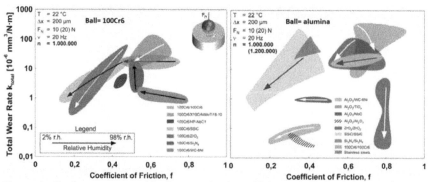

Figure 4. Total wear rates with the associated coefficient of friction under unlubricated oscillation (fretting) conditions and influence of the relative humidity

-Dry oscillation

The tribological profile (K_v versus CoF) is displayed in Figure 4 for ceramic and steel counterbodies. The arrows indicate an increasing relative humidity. Against a polished alumina ball ($\varnothing$= 10 mm), the tribological profiles of steel and the ceramic samples including WC-6Ni and binderless HP-NbC are sensitive to relative humidity. Against a polished 100Cr6 (SAE E52100) ball ($\varnothing$= 10 mm) only HP-NbC was found to be insensitive to humidity. Thus, tribo-oxidation seems not to affect the tribological characteristics of HP-NbC when oscillating against ball bearing steel 100Cr6. In general and in particular in comparison to WC-6Ni, the wear resistance of HP-NbC under dry oscillation is good having K_v value of 10^{-6} mm³/Nm.

A direct comparison between WC-6Ni and HP-NbC1 is difficult as nickel binder in WC-6Ni (6,4 wt.-% Ni + 0,5 wt.-% Co) has a beneficial role in terms of tribo-oxidation (formation of double oxides [8]) and the WC grain size being around 2 μm) is much smaller than that of HP-NbC. Furthermore, the coefficient of friction for both WC-6Ni tribo-couples depends much stronger on the relative humidity than that for HP-NbC1 tribo-couples.

CONCLUSIONS

The wear resistance of the present HP-NbC can easily compete with ceramics and hard metals, thus qualifying NbC for the group of tribological materials with enhanced wear resistance. The room temperature wear rate of HP-NbC increased from <10^{-8} mm³/Nm at low sliding speeds towards 10^{-6} mm³/Nm when the sliding speed was raised to 7,5 m/s. The wear rate at 400°C remained generally below 10^{-6} mm³/Nm regardless of the applied sliding speed. Under dry oscillation, the wear resistance of HP-NbC is insensitive to relative humidity for bearing steel (100Cr6=SAE E52100) as well as alumina counterbodies. Future optimizations such as the elimination of grain porosity and grain refinement as well as the use of specific sintering aids and metallic binders will very likely result in further improvements of the wear resistance of NbC-based materials.

REFERENCES

[1] H. Mohrbacher and Q. Zhai, Niobium alloying in grey cast iron for vehicle brake discs, Proceeding of Materials Science & Technology 2011, p. 434-445
[2] T. Nylén, Niobium in Cast Iron, Proceedings of the International Symposium Niobium 2001, TMS (2001), 1063-1080
[3] M. Woydt and K.-H. Habig, High temperature tribology of ceramics, Tribology International, 1989, Vol. 22, No. 22, p. 75-88
[4] L.-M. Berger, C. C. Stahr, S. Saaro, S. Thiele, M. Woydt and N. Kelling, Dry sliding up to 7.5 m/s and 800°C of thermally sprayed coatings of the TiO₂–Cr₂O₃ system and (Ti,Mo)(C,N)-Ni(Co), WEAR 267 (2009) 954-964
[5] D. Klaffke, Fretting Wear of Ceramics, Tribology International, 1989, Vol. 22, No. 2, p. 89-101
[6] L.-M. Berger, S. Saaro and M. Woydt, (WC-(W,Cr)₂C-Ni— the unknown hardmetal coating, Thermal Spray Bulletin, (bilingual), 1/08, p. 37-40
[7] A. Skopp and M. Woydt, Ceramic and Ceramic Composite Materials with Improved Friction and Wear Properties, Tribology Transactions, Vol. 38(2),1995, p. 233-242
[8] M. Woydt, A. Skopp, I. Dörfel and K. Wittke, Wear engineering oxides/Anti-wear oxides, Tribology Transactions Vol. 42, 1999, No. 1, p. 21-31 or in WEAR 218 (1998) 84-9

TRIBOLOGICAL PROPERTIES OF ALUMINA/ZIRCONIA COMPOSITES WITH AND WITHOUT h-BN PHASES

Liang Xue[1] and Gary L. Doll[2]
[1] The Timken Company, Technology Center, Canton, OH, USA
[2] University of Akron, Engineering Research Center, Akron, OH, USA

ABSTRACT

The tribological performance of alumina/zirconia composites (alumina with 10 vol% zirconia) with and without h-BN added to the material has been examined. Whereas the baseline composite (without h-BN) was conventionally sintered, h-BN containing composites were sintered by Spark Plasma Sintering. Friction and wear tests were conducted in unlubricated, reciprocating sliding. Testing of self-mated pairs revealed that only composites with low concentrations of h-BN exhibited low friction coefficients. When mated against steel, all compositions had relatively high friction coefficients and had noticeable wear, especially with higher concentrations of h-BN. The tribological behavior of the composites with higher amounts of h-BN appeared to be greatly influenced by residual porosity and h-BN particle pull-out.

INTRODUCTION

Alumina/zirconia composites possess some desirable material properties, e.g., high hardness, low density, corrosion resistance, and high temperature capability, which may be advantageous for use as bearing components [1]. These composites could be low cost alternatives to silicon nitride since they are a lower-cost material and can be fabricated without expensive HIPing processes.

The incorporation of hexagonal boron nitride (h-BN) particles as a third phase material into the alumina/zirconia composite may be tribologically beneficial. h-BN is well known as a good solid lubricant at both low and high temperatures (up to 900 °C), even in an oxidizing atmosphere [2]. h-BN as a lubricant is particularly useful when the electrical conductivity or chemical reactivity of graphite would be problematic. Another advantage of h-BN over graphite is that its lubricity does not require water or gas molecules trapped between the layers. Therefore, h-BN lubricants can be used even in vacuum, e.g. in space applications. h-BN phases have been reported to significantly reduce the friction coefficient and wear of silicon nitride ceramics [3]. h-BN is both thermally and chemically stable at the alumina/zirconia composite sintering/processing temperature that could approach 1700°C. Although the self-lubricating function of many solid oxide lubricants (MO_x; M=Cu, Ti, Mg, Zn, Mn) has been studied in alumina, zirconia, and alumina/zirconia composite [4-6], there are very few published reports on the use of h-BN as a solid lubricant in alumina/zirconia composites. Kim and Lee have reported the friction and wear behaviors of alumina/zirconia nanocomposites containing h-BN against an alumina ball in ball-on-plate tests [7]. In their study, the h-BN composite shows a lower friction coefficient at room temperature but a higher wear rate at high temperatures, and that the friction does not change much with the amount of h-BN (2 to 10 wt%).

In this exploratory study, the tribological properties of alumina/zirconia composites (alumina with 10 vol% zirconia) with and without h-BN added to the material have been investigated for potential self-lubricating capability.

EXPERIMENTAL PROCEDURE

Alumina with 10 vol.% zirconia was chosen as the baseline material since it is a low cost material and relatively easy to synthesize. Starting materials were commercially available submicron alumina powder (P172SB, Rio Tinto Alcan), submicron yttria-partially stabilized zirconia powder (3Y-TZP, Tosoh), and nano-sized h-BN powder (NX1, Momentive Performance Materials). Appropriate amounts of powders were mixed by wet-milling in a polyethylene bottle, using zirconia cylinders as milling media, in either water or isoproponal alcohol. When water was used, ~1.5wt% Darvan 821A (ammonium phosphate from R.T. Vanderbilt) was added as dispersant. After drying, the powder was sieved through a 100 mesh sieve.

For the baseline composite without h-BN, green pellets of the powder with diameters of ½" and 1-1/4" were prepared by die pressing. The pellets were then sintered in air at 1550°C for 2 hours. After sintering, the pellets reached a relative density as high as 99.8%.

The addition of h-BN to the composites would impede the sintering. The h-BN would oxidize in air at the sintering temperature and require an inert sintering atmosphere. Consequently, the h-BN added composites were sintered by SPS (Spark Plasma Sintering) in vacuum/argon.

Powders of two different h-BN contents, i.e., 5 vol.% and 20 vol.% were used to synthesize the composites with h-BN. The powders were wet-mixed, dried, and then calcined in air at 550°C for 2 hours to remove any organic residues (e.g., Darvan 821A) prior to sintering. The calcined powders were then sintered by SPS in a 40 mm diameter graphite die in vacuum during heating at a rate of 100°C/min. and then in argon at temperatures ranging from 1500°C–1680°C and a pressure of 35–55 MPa for 6 to 15 min.

Hardness and fracture toughness were obtained by Vickers indentation on polished surfaces of cross-sectioned samples, with an indentation load of 15 kg. The fracture toughness values were calculated according to the equation reported by Lawn et al [8]:

$$Kc = 0.0016(E/H)^{1/2}(P/c^{3/2}) \tag{1}$$

where in Kc is the fracture toughness, E represents Young's modulus, H is the hardness, P stands for the peak load on the Vickers indenter, and c gives the crack length, measured from the center of the indent to the crack tip. Polished sample cross-sections were examined by scanning electron microscopy (SEM) and analyzed by EDS. Some samples were thermal etched at 1150°C in nitrogen for 30 min. prior to SEM examination.

Friction and wear tests were conducted on a Plint TE-77 machine in reciprocal sliding mode under dry conditions at room temperature. A schematic and a picture of the TE-77 machine are shown in Figure 1 below.

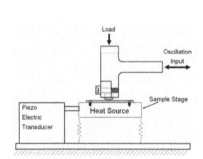

Figure 1. Schematic and picture of the TE-77 machine for tribological tests.

The following testing parameters were used: 20 N normal load, 1.8 mm stroke length, 5 Hz frequency, and 2000-5000 cycles duration. The samples were tested either self-mated, i.e., a rectangular composite piece (~3mm x ~3mm of contact area) on a flat composite plate, or against steel, where 7 mm diameter 52100 balls were in contact with the flat composite plates.

RESULTS AND DISCUSSION

SEM Microstructure
Figure 2 displays the microstructure of a 5% h-BN-alumina/zirconia composite sample SPS-sintered to 1550°C at 50 MPa for 20 min., which reached a relative density of over 99.1%. The sample exhibited a reasonably uniform distribution of the zirconia and h-BN phases.

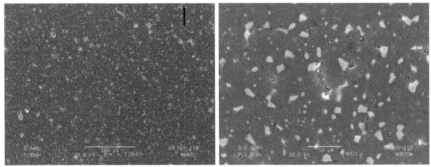

Figure 2. SEM images of 5% h-BN-alumina/zirconia composite.
Left: low magnification. Right: high magnification.

Figure 3 shows the microstructure of a 20% h-BN-alumina/zirconia composite sample SPS-sintered to 1680°C at 55 MPa for 10 min. that reached a relative density of 98.3%. It also shows a reasonably uniform distribution of zirconia and h-BN phases.

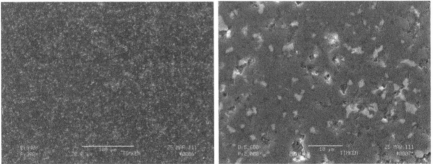

Figure 3. SEM images of 20% h-BN-alumina/zirconia composite with relatively high density.
Left: low magnification. Right: high magnification.

Some of the surface pits seen in Figure 3 are assumed to be due to the pull-out of the h-BN particles during sample preparation and polishing, resulting in a much higher apparent surface porosity and a rougher surface. h-BN pull-out during sample polishing could possibly occur even in fully dense composites, because of the extreme softness of the h-BN as compared to alumina in the surrounding matrix.

EDS Analysis

EDS analyses were conducted in conjunction with SEM to verify the presence of BN in the BN-added composite samples.

Figure 4 shows the locations where a 5% h-BN-alumina/zirconia composite sample was analyzed by EDS. Locations 1 and 2 are thought to be grain boundaries and grain junctions where BN particles are likely concentrated. The corresponding EDS results are listed in Table 1 below.

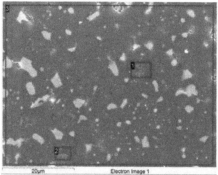

Figure 4. EDS analysis locations of 5% BN-alumina/zirconia composite.

Table 1. EDS analysis results in weight% for locations in Fig. 4.

Spectrum	B	N	O	Al	Zr	Total
1	12.47	3.40	40.14	43.33	0.66	100.00
2	14.38	2.55	37.27	41.80	4.00	100.00
3	4.63	2.89	41.79	41.43	9.26	100.00

The EDS data, without reference standards, are semi-quantitative in nature. Even though the absolute values are off, relative comparison among the different locations can be made. Locations 1 & 2 (with respective Spectra 1 & 2) had a boron content about 3 times as high as the average of the whole area as shown in Spectrum 3, confirming the presence of BN in these grain boundary and grain junction areas.

EDS analysis was also performed on a 20% BN-alumina/zirconia composite sample. A high-magnification SEM photograph, Figure 5, shows the locations of the EDS analyses. The Dark phases in locations 1 & 2 were thought likely to be BN grains, the bright phase at location 3 was probably zirconia, and the gray grain at location 4 should be alumina. The EDS analysis results are listed in Table 2.

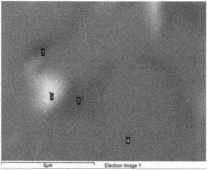

Figure 5. EDS analysis locations of 20% BN-alumina/zirconia composite.

Table 2. EDS analysis results in weight% for locations in Fig. 5.

Spectrum	B	N	O	Al	Zr	Total
1	29.45	15.96	24.21	25.26	5.12	100.00
2	45.13	18.77	15.74	18.15	2.21	100.00
3	15.14	9.17	29.29	17.19	29.21	100.00
4			45.53	56.47		100.00

The EDS data in Table 2 show that locations 1 & 2 were indeed high in boron concentration (much higher than that of 5% BN sample shown in Table 1, as expected), location 3 was rich in Zr, and location 4 was alumina, confirming the presence of these materials in the composite.

Hardness and Fracture Toughness

The material properties of baseline alumina/zirconia (AZ), 5% BN-alumina/zirconia (AZ-5v%BN), and 20% BN-alumina/zirconia (AZ-20v%BN) are listed in Table 3. The values of elastic modulus are calculated theoretically from individual material data. The hardness and fracture toughness are measured and calculated from Vickers indentations.

As can be seen from the table, the material hardness and elastic modulus decrease with the increase of BN content in the composite. Because hexagonal boron nitride is a very soft material, at 20 vol.% addition, the hardness of the composite is reduced to almost half that of the baseline alumina/zirconia. On the other hand, the fracture toughness of the composite increases with the increase of the BN content in the composite, due to the enhanced toughening mechanism of crack deflection by the additional second phase [9]. The toughness of 20% BN-alumina/zirconia is about 50% higher than the baseline composite.

Table 3. Material properties of the composite materials.

Material Properties	AZ	AZ-5v%BN	AZ-20v%BN
Elastic Modulus (GPa)	372	355	305
Hardness (GPa)	17.0 ± 0.6	13.4 ± 0.3	9.2 ± 0.3
Toughness (MPam$^{1/2}$)	4.0 ± 0.2	4.5 ± 0.2	6.0 ± 0.2

The hardness of the baseline composite (~17 GPa) is higher than that of silicon nitride (~15 GPa), but its fracture toughness, which is more important for rolling contact fatigue life for example, is lower than that (~6 MPam$^{1/2}$ for good quality materials) of silicon nitride [10].

Tribological Tests - Friction and Wear Properties

Tribological tests were performed on composite samples, either sliding against itself (self-mated) or against a steel ball. Figure 6 shows the self-mated coefficient of friction vs. test cycles for the earlier batch composites where AZ stands for baseline alumina/zirconia and AZB stands for 5 vol% h-BN added alumina/zirconia. These earlier batch AZB composites had over 2% residual porosity.

As can be seen in Figure 6, after an initial wear-in period of about 1500 cycles, the friction coefficient of both materials somewhat stabilized in the 04 - 0.5 range, with the 5 vol% BN composite having a slightly higher friction coefficient.

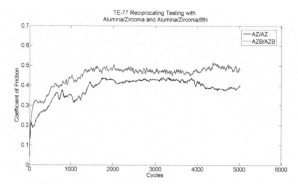

Figure 6. Tribological tests of self-mated composites of earlier batches.

The self-mated dry sliding friction data on alumina/zirconia composites are not available in the literature. However, the 0.4 ~ 0.45 coefficient of friction displayed by the baseline alumina/zirconia composite (AZ) in this work is close to that (~0.45) reported in the literature where alumina/zirconia composite (of the same composition as this work) slid against alumina [1, 4]. This value is also close to, albeit slightly lower than, that of 0.52 reported on sliding of alumina/zirconia composite against zirconia (TZ-3Y) [11].

The slightly higher coefficient of friction displayed by the 5 vol% BN composite (AZB) was unexpected based on the theory that h-BN would function as a solid lubricant to reduce the friction. However, it is likely that the existence of over 2% residual porosity in the sample impacted its friction performance due to the fact its polished surface was much rougher than the baseline composite.

Figure 7 shows the measured coefficient of friction from the composite materials mated with steel. Again, after an initial wear-in period of about 1500 cycles, the friction coefficient of both materials stabilized, though at higher values than shown in Figure 6. The coefficient of friction of baseline composite was ~0.6, while that of the 5 vol% BN composite was slightly higher at ~0.65.

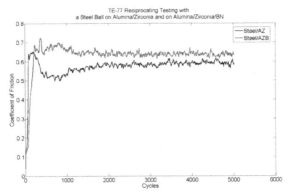

Figure 7. Tribological tests of earlier batch composites against steel.

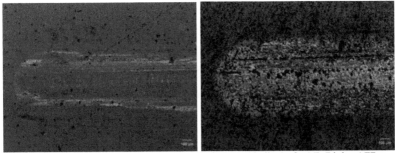

Figure 8. Wear tracks of the composites against steel balls. Left: AZ; Right: AZB.

The wear tracks produced in the reciprocating sliding testing of these two samples are shown in Figure 8. Both wear tracks show an increased roughness of the h-BN containing composite material as well as some arc tensile cracks within the wear track on the alumina-zirconia composite. In comparison, the h-BN composite yielded significantly more wear as compared to the alumina-zirconia composite. A portion of the increased wear on the h-BN containing material may be due to the ~5% lower elastic modulus of the AZB sample, which would generate a slightly wider wear track as compared to the alumina-zirconia composite. The reduced modulus coupled with an increased porosity and roughness may be responsible for the larger friction and greater wear of this material. The ~0.6 coefficient of friction of the baseline composite against steel is about the same as that of a silicon nitride/steel pair, where values of between 0.56 and 0.61 have been reported [12].

Figure 9 shows the self-mated coefficients of friction vs. test cycles for improved h-BN containing batch composites where AZB2-1 and AZB2-2 were 5 vol% BN composite samples and AZB3-1 and AZB3-3 were 20 vol% BN composite samples. AZB2-1 and AZB2-2 all reached a relatively high sintering density of over 99%. However, AZB3-1 and AZB3-3, due to high BN content, still had residual porosities of close to 2%.

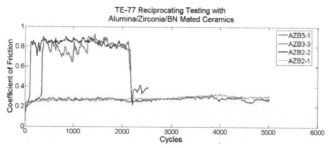

Figure 9. Measured friction from tribological tests of self-mated BN-added composites of improved batches.

Figure 9 shows that AZB2-1 and AZB2-2, which had 5 vol% BN, exhibited low self-mated friction, with coefficients of friction about 0.27. In contrast, AZB3-1 and AZB3-3, which had 20 vol% BN, showed a larger coefficient of friction of about 0.8, which may again be due to the increased porosity in the 20% vol samples.

Whereas the coefficient of friction of h-BN as a solid lubricant was reported to range from 0.15 to 0.70 [13], the value of ~0.27 obtained in this study (for the 5% BN composite) tends towards the lower end of the range.

Figure 10 (a) shows photo of 5 vol% BN composite flat plate sample after the self-mated tribological tests. The lack of any visible wear scars suggests that the addition of h-BN in this material reduced the friction and the wear between the self-mated composite materials. In contrast, the 20 vol% BN composite exhibited significant wear during the tests, as shown in Figure 10 (b).

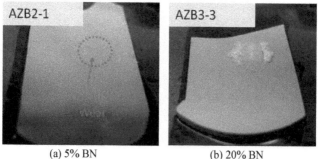

(a) 5% BN   (b) 20% BN

Figure 10. Photos of samples after self-mated tribological tests.

The measured friction from the tribological testing of these two composites mated with steel balls are plotted in Figure 11. They all show relatively high friction values. One of the 5% BN samples (AZB2-2) had ~0.65 coefficient of friction, which was about the same as that of the previous sample tested as shown in Figure 7. The other 5% BN sample (AZB2-1) shows a higher value of ~0.9. The two 20% BN samples (AZB3-1 and AZB3-3) show even higher friction coefficient value in the 0.8 to 1.0 range.

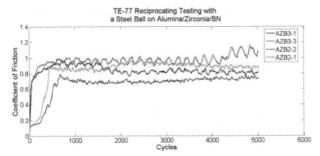

Figure 11. Tribological tests of improved batches of BN composites against steel.

Figure 12 shows the wear tracks of the two 5 vol% BN composite flat plates after tribological tests against steel balls. The corresponding contact/wear surfaces of the steel balls are shown in Figure 13. The wear tracks on both composite samples are quite similar, and the corresponding wear marks (appeared as a black circle) on the steel balls are about the same. The overall width of the wear tracks on both samples is about 0.96 mm, which is very close to the 0.95 mm diameter wear marks measured of the steel balls. However, the deep wear grooves at the center of the wear tracks were much narrower (only about 0.32 mm in width for both samples). The wear tracks were ~2.7 mm in length, which can be closely approximated by adding the two radii of the steel ball wear marks (=0.95 mm) to the 1.8 mm stroke length of the ball.

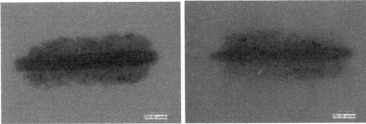

Figure 12. Wear tracks of 5% BN composites against steel balls. Left: AZB2-1; Right: AZB2-2.

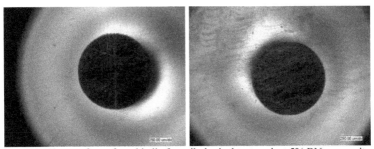

Figure 13. Wear surface of steel ball after tribological test against 5% BN composites.
Left: on AZB2-1; Right: on AZB2-2.

The wear tracks of the two 20 vol% BN composite flat plates (AZB3-1 and AZB3-3) after tribological tests against steel balls are shown in Figure 14, and the corresponding contact/wear surfaces of the steel balls are shown in Figure 15. Wear tracks of these two 20% BN composites are much wider (~1.6 mm and ~2.0 mm, respectively) than that (~0.96 mm) of the two 5% BN samples shown in Figure 12. In addition, the corresponding dimensions of the wear marks on the steel balls once again correspond with the wear track width of the composite sample. Furthermore, the wear scar of the steel ball generated from the AZB3-3 sample appears to have an oval shape, with the long axis having about the same length (~2.0 mm) as that of the wear track width.

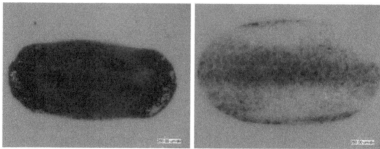

Figure 14. Wear tracks of 20% BN composite flat plate against steel balls.
Left: AZB3-1; Right: AZB3-3.

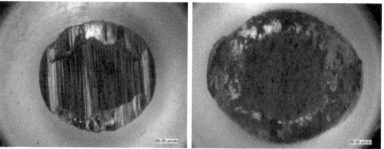

Figure 15. Wear surface of steel ball after tribological test against 20% BN composites.
Left: on AZB3-1; Right: on AZB3-3.

Based on the test results, it is conceivable that the presence of ~ 2% residual porosity in the 20% BN composite samples may have negatively impacted their wear behavior. In addition, the pull-out of the BN particle agglomerates during sample preparation and polishing may have increased the surface roughness and reduced the h-BN content at the tribological interface (more so for the 20% BN ones). These factors may explain some of the differences and scatter in their tribological performance.

The test results indicate that h-BN in the composites will reduce the friction between mated composite surfaces, but may not function well as a solid lubricant against steel in dry sliding conditions. This attribute may limit the composite's use in applications with predominantly sliding contacts, such as in journal bearings.

In this exploratory work, we primarily focused on the friction behavior of the composites and observed the wear behavior in only a qualitative way. Therefore, no direct comparison in terms of wear rates with other tribological materials can be made. Quantitative analysis of the wear rate will be performed and reported at a later date.

SUMMARY

The baseline alumina/zirconia composite (alumina with 10 vol% zirconia) can be pressurelessly sintered to 99.8% density. The potential self-lubricating capability of these materials was explored by introducing 5 and 20 vol% h-BN to the composite. These h-BN containing composites were sintered by SPS to high densities but still with close to 2% residual porosity for the 20 vol% BN-added material.

Friction and wear tests conducted under dry conditions in reciprocal sliding mode showed that the self-mated friction coefficient of baseline alumina/zirconia composite is ~0.45 and agrees with what has been reported in the literature. Only the self-mated 5 vol% h-BN exhibited low friction values (<0.3). The self-mated 20 vol% h-BN yielded a very large friction coefficient of ~0.8. When sliding against steel, all compositions had friction coefficients ~0.6 or greater and all had measurable wear, especially with 20 vol% h-BN. However, the wear behaviors were observed only qualitatively.

The poorer performance of the 20% h-BN containing material may have been a result of relatively high residual porosity. BN particle pull-out during the sample preparation and polishing is believed to be the mechanism that contributed to the relatively high friction and wear of the BN composites. It appears that residual porosity may also have a significant impact on the friction and wear behavior of BN-alumina/zirconia composites.

These preliminary tribological test results suggest that the h-BN composites may possibly function well in all-composite bearings. However, they may not function well against steel in dry sliding conditions. Therefore they would appear to have limited use in applications with predominantly sliding contacts such as in journal bearings. Further quantitative studies and comparisons with other tribological materials are recommended.

ACKNOWLEDGEMENTS

The authors would like to acknowledge the assistance and contributions of Carl Hager, Matt Boyle, Jerry Richter, Bob Pendergrass, Michelle Petraroli, Mary Theiss, and Mike Byrne of the Timken Company. They also wish to express gratitude to the Center for Innovative Sintered Products at Penn State University for the use of their equipment including SPS.

REFERENCES

1.   B. Kerkwijk, et al., "Tribological Properties of Nanoscale Alumina-Zirconia Composites," Wear 225-229, 1293-1302 (1999).
2.   J. Greim, K. A. Schwetz, "Boron Carbide, Boron Nitride, and Metal Borides", in Ullmann's Encyclopedia of Industrial Chemistry. Wiley-VCH: Weinheim (2005).
3.   W. Chen, et al., "Tribological Characteristics of Si3N4-hBN Ceramic Materials Sliding against Stainless Steel without Lubrication," Wear 269, 241-248 (2010).
4.   A. K. Dey and K Biswas, "Dry Sliding Wear of Zirconia-Toughened Alumina with Different Metal Oxide Additives," Ceramic Int'l 35, 997-1002 (2009).
5.   B. Kerkwijk, et al., "Friction Behavior of Solid Oxide Lubricants as Second Phase in α-Al_2O_3 and Stabilized ZrO_2 Composites," Wear 256, 182-189 (2004).
6.   S. Ran, et al., "Dry-Sliding Self-Lubricating Ceramics: CuO Doped 3Y-TZP," Wear 267, 1696-1701 (2009).
7.   S. Kim and S. Lee, "Fiction and Wear Behaviors of Al2O3-15wt% ZrO2-Solid Lubricants Nanocomposites at Temperature up to 600 C". Paper ICACC-S1-085-2011 presented at 35th Int'l Conf. & Exp. on Adv. Ceramics & Composites, Daytona Beach, FL, USA, Jan. 23-28, 2011.
8.   B. R. Lawn, A. G. Evans, and D. B. Marshall, "Elastic/Plastic Indentation Damage in Ceramics: The Median/Radial Crack System," J. Am. Ceram. Soc. 63 [9], 574-581 (1980).
9.   K. T. Faber and A. G. Evans, "Crack Deflection Processes-I, Theory, and II, Experiment," Acta. Metal. 31 [4], 565-584 (1983).
10.   L. Xue and G. Doll, "Fatigue Lives and Material Properties of Silicon Nitride Balls for Hybrid Bearing Applications," pp 259-271 in Fatigue of Materials Advances and Emergences in Understanding, Edited by T.S. Srivatsan and M. Ashraf Imam. TMS, 2010.
11.   Y. J. He, et al., "Effects of a Second Phase on the Tribological Properties of Al_2O_3 and ZrO_2 Ceramics", Wear 210, 178-187 (1997).
12.   P. Reis, et al., "Friction and Wear Behaviour of Beta-Silicon Nitride-Steel Couples Under Unlubricated Conditions," REDORBIT NEWS (2006).
www.redorbit.com/news/display/?id=446144. Published: 2006/03/28.
13.   M K Impex Canada, "Physical Properties of Hexagonal Boron Nitride," www.lowerfriction.com.

Author Index

Author Index

Taylor, B., 111
Teymouri, M., 179
Teyssandier, F., 79

Vandeperre, L. J., 69
Vick, M., 69
Vignoles, G. L., 137
Voggenreiter, H., 11

Weisbecker, P., 79
Welch, S. T., 111
Wen, W., 57
Woydt, M., 189

Wu, L., 89, 95

Xiao, J., 57
Xue, L., 195

Yang, J.-M., 123
Yuan, H., 57
Yuan, W., 155

Zhang, H., 57
Zheng, J. Q., 101
Zhu, H., 155
Zhu, Q., 155